Final Theory of Light

& *Finding Extraterrestrials*

Russell Eaton

Copyright

Copyright © 2024 Russell Eaton. All rights reserved. No part of this book may be used or reproduced in any form whatsoever without permission except in the case of brief quotations in articles or reviews.

Title: Final Theory of Light
Subtitle: & Finding Extraterrestrials
Author: Russell Eaton
Publisher: DeliveredOnline.com
ISBN (eBook): 978-1-903339-78-7
ISBN (paperback): 978-1-903339-02-2
This paperback edition published 22nd September, 2025

For any queries please contact the publisher:
Email: mailto@deliveredonline.com
Website: www.deliveredonline.com

Colour images for this publication
The ebook is in colour, but this depends on your eReader model. This physical paperback book is not printed in colour so as to keep the price as low as possible. If you wish to see any images in colour please go here in your browser to instantly download all the book's images in colour, free of charge:
https://drive.google.com/file/d/1FZvdW6DisJtGweeV4dvgZotVtRh5MVLr/view

*

Content

Copyright 3
Preface 7
Introduction 9
A brief overview of light 13
What is light made of? 15
How is light created? 17
What is incident light? 21
What is the journey-time of light? 31
What is the frequency of light? 37
What is the amplitude of light? 43
How does light move? 45
Why is the constant speed of light constant?
.... 51
What is the energy of light? 55
What is the big misunderstanding of light?. 61
Insurmountable contradiction 65
The brief history giving rise to the misunderstanding 69
Double slit experiments 71
Contemporary wave-theory of light 77

Do light-waves exist? 93
What is the invisible spectrum of light? 97
How do we see colours? 103
How do we see a specific colour? 119
How does a prism work? 131
Is the quantum theory of light correct? 147
Eaton's Constant 153
How is light-energy measured? 159
What is the brightness of light? 163
Why can't the speed of light change? 165
Does light have mass? 173
Can light ever bend? 177
Does light carry information? 187
Virtual Video Camera 191
Finding Extraterrestrials 199
Message from the author 211
Final Theory of Everything 213
Author Bio 217

*

Preface

We must always endeavour to eradicate bigotry and prejudice from science, and we must always be on guard when such pernicious influences come knocking at the door.

Russell Eaton

Author's Apology
The author apologies for any hurt feelings or dismay caused by the revelations in this book. It is understandable that talk of Einstenian relativity or quantum physics being spurious may upset some people in the scientific community whose career and credibility depend on the veracity of such science. No disrespect is intended by the author.

To see images in colour please see copyright page for information

*

Introduction

Light is a truly wondrous phenomenon of nature. In physics light is a well-studied subject, yet there are many misconceptions concerning light's fundamental nature. Some mysteries about light continue to puzzle scientists even today in the 21st Century. But now, in the *Final Theory of Light,* these puzzles can be resolved and revealed for the first time. Here are just some of the mysteries of light that are resolved in the book:

* Why the speed of light is in fact always constant, even when moving inside a medium.

* Exactly how light carries information to Earth from distant parts of the universe.

* Why all photons in the universe are identical in every respect, thus busting a widespread misconception.

* The spurious nature of the so-called 'quantum theory of light'.

* The wavelength of light determines which colours we see, but why? What exactly determines the length of wavelengths? This mystery is fully resolved.

* Is light both a wave and a particle? The mystery of light's duality finally put to rest.

* The famous double-slit experiments that misled the world.

* Why light can never bend, bounce or reflect off anything.

(And much more)

This book gives you a fundamental understanding of the nature of light as never before, and you will learn about a *virtual video camera* that is destined to revolutionise humanity's exploration of the Universe.

One day soon humans will be able to obtain full video recordings (with sound and colour) of planets and stars, as if we had put a physical video camera on the actual surface of a planet or star. We will be able to do this from Earth instantaneously, distance no barrier. This book reveals a never-published-before method that shows exactly how to discover extraterrestrial life and dramatically alter our knowledge of the cosmos.

Particle physics has not seen progress since the 1970s when the standard model of particle physics was completed. Ever since then, the theories used to describe observations in physics have remained unchanged. Little by little the standard model of particle physics has become more and more outdated and inconsistent.

As a result, millions of physics students are today going down blind alleys and rabbit holes full of misguided concepts. This in turn leads to blighted careers, and a falling out in the pursuit of science.

As pointed out by Charlie Wood and many others *(Fundamental physics is in a crisis, Quanta Magazine, 12 August 2024)*, scientists are increasingly saying that particle physics is facing a nightmare scenario with many researchers looking for a new direction in physics.

A monumental shift in a new direction in particle physics is very overdue. This book provides that major shift, setting particle physics in a new direction and the prospect of many new exciting discoveries.

The *Final Theory of Light & Finding Extraterrestrials* is for everybody to read and enjoy whatever your background or expertise. The book is available in English (Final Theory of Light & Finding Extraterrestrials) or in Spanish (Teoría Final de la Luz y la Búsqueda de Extraterrestres).

*

A brief overview of light

When we see light, we are seeing streams of photons coming into our eyes. When light is created it radiates out in all directions, in straight lines. So when some of those straight lines of light go to our eyes, this is how we see things. Light always moves at the same speed (about 300 million metres per second); it doesn't slow down or speed up, it doesn't bounce off anything or curve in any way, and it continues moving indefinitely unless and until something gets in its way. If you have heard that light reflects or bounces off things, or that it bends, none of this is correct. You are urged to discover the truly wondrous nature of light by reading on.

For clarity, the information in this book is mostly presented in the form of questions and answers.

*

What is light made of?

Light is entirely made of photons. And photons are entirely made of oscillating electromagnetism. So light consists of streams of separate photons moving in straight lines in all directions. It is widely, though mistakenly, believed that a photon is an elementary and indivisible particle of light. In fact a photon is simply a convenient word for referring to an oscillating packet of energy that is self-contained and separate from other photons. So a photon is a little self-contained oscillating electromagnetic field. It started oscillating as soon as the light was created or emitted, and at any given moment the photons that make up light would have been oscillating many trillions of times without ever running out of energy.

It is thought that a single electromagnetic oscillation represents the elementary particle of light, this being the smallest known quantum of energy in the Universe. Each electromagnetic oscillation represents the total energy of a photon, but such oscillations are not cumulative. The same basic energy of one oscillation remains the total energy of a photon however many trillions of times it may have oscillated.

When we say that light is made of streams of photons, we are in fact saying that light is made of streams or groups of separate little oscillating fields of electromagnetism. By 'separate' we mean the photons are not joined up or coupled in some way, yet nevertheless most photons travel together as a stream because that is how they were emitted. Photons never travel as part of a single wave (a single energy field of multiple photons).

Evidence that photons are not joined up when travelling as a stream is the simple fact that once emitted, the many streams of photons expand (radiate) outward in all directions, albeit always in straight lines. So when the photons expand outward in all directions they expand as trillions of separate photons, mostly moving with other separate photons in many streams.

*

How is light created?

When atoms of an object are heated up or made to move more quickly, they generate electromagnetism. So light is created from the excitation of atoms in an object such as the sun, a light bulb, a flame, a torch, etc. The hotter the atoms in an object the greater the emission of light in the form of photons.

Technically, the excitation of an atom also excites the electrons of said atom. This results in the emission of photons from the electrons (all photons originate from electrons). When this happens, each electron emits only one photon at a time that shoots out of the atom on its own. Of course, this means that millions of electrons in many atoms can be emitting photons in all directions, as a growing sphere of light.

"This new paradigm of photon production enables the production and emission of photons at rates ... in the hundreds of trillions per second. This has enabled the explanation of every branch of physics, not through the inaccurate, and esoteric implementation of wave-functions but through the easily understood and calculable classical physics paradigms". Source: Dilip D James, Electricity and Radio waves according to Augmented Newtonian Dynamics,International Journal of Science and Research 13(12):1222-1228.

When an electron creates a photon it emits a packet of electromagnetic energy that we call a photon, and the kinetic energy of the electron helps it to send the photon flying off at the speed of light. The speed of light is set by the universal rate of electromagnetic oscillations of a photon.

"A single atom, by its nature, can only emit one photon at a time" (source: Professor Gerhard Rempe, Max Planck Institute of Quantum Optics, Germany, mpg.de, 2007).

There are many ways to create light in today's science. And we experience light in many ways: a light bulb, a matchlight, a torch, a flame, sunlight, starlight, and so on. Also, the intensity of light varies, ranging from microwaves and infrared, to the visible light that we see. Everything that we see around us is made possible to see by the existence of light coming from a 'natural source' such as sunlight, or a 'synthetic source' such as a light bulb.

The maximum (i.e. constant) speed of light at 'c' is set by electrons. Whenever electrons emit photons, the photons are always emitted with the same amount of electromagnetic energy everywhere in the Universe. Why so? Because when electrons move from any orbit to the next orbit *closer* to the nucleus of an atom, the electron gives off the same amount of electromagnetic energy in the form of a photon.

> **Every single photon in the universe is entirely created by an electron, and every photon is created equal**

All photons are born with the same amount of oscillating electromagnetic energy, thus setting their constant speed at 'c' (the speed of light).

What is incident light?

Understanding the meaning of incident light is fundamental to understanding the nature of light. So-called 'incident light' refers to light that is not coming directly to you from the original source that *created* the light. Everything that we see around us in our daily lives is made possible to see because of incident light.

As light can never bounce or be reflected, what happens instead is that the light is absorbed into things around you, and then brand-new incident light is emitted in its place. You can think of incident light as replacement light because all incident light is light that replaces directly-created light, i.e. replaces directly-created light that has been absorbed. So incident light is in effect a replacement light.

When light is absorbed into a material, medium or object it is gone forever by being converted to heat and other forms of energy.

For example, when daylight hits a red car, the photons of the daylight are absorbed into the atoms located below the red paint and then new incident photons are emitted. Those incident photons travel to our eyes and we see a red car. The incident photons are not somehow encoded with the colour red or the image of a car, so how do we end up seeing a red car?

This is what happens. The incident light comes out of the red car as streams of incident photons and go in all directions in straight lines. But those streams of incident photons have a journey-time that is a little bit slower than the journey-time of the speed of light. Why so? Because although each incident photon moves at the speed of light, a small time-interval occurs between each photon streaming out of the car. This time-interval is caused by the time it takes the electrons in the atoms of the car to absorb photons and then emit newly created (different) photons.

This means that the stream of incident light coming from the red car to your eyes has a journey-time that is a little slower than the normal speed of light. To be clear, each photon as such does not slow down, but the stream as a whole slows down. Here's a somewhat more technical description of incident light:

The amount by which light is *slowed down* as a result of absorption and emission is called the refractive index. and the process itself of absorption & emission is called attenuation. Light has no existence except as a photon travelling at speed 'c' (the constant speed of light). Light is absorbed by the first layer or so of atoms of a material or medium it comes upon. The incident light is then 'reconstructed' in accordance with the characteristics of the atoms of the material receiving the light. Some materials take longer to attenuate light.

More specifically, the incoming photons that hit electrons will over-energise such electrons and make them unstable. When this happens, the electrons are compelled to release their excess energy in the form of newly created photons.

The absorption and emission of photons takes the electrons a moment of time to accomplish. This puts a particular distance between each emitted photon. This distance determines the journey-time of light for any given light ray or group of photons. The greater the distance between each moving photon, the greater the journey-time of that whole light ray. So although any individual photon always moves at the constant speed of light, the journey-time of a given group of photons can vary. More about this throughout the book.

The key point here is that all photons in the universe are identical and every photon carries the same energy. When such energy is absorbed by an electron, the electron will release a new photon with exactly the same energy as the amount absorbed.

There are many studies showing this to be so, as in the following example:

"When a photon is absorbed into an electron, the electron is energised making it change levels. In doing so the electrons in the atom emit photons. The photon is emitted with the electron moving from a higher energy level to a lower energy level. The energy of the photon emitted has the exact same energy as the absorbed photon, i.e. the electron loses the exact energy received by moving to its lower

energy level" (source: Photon Emission, Dept of Physics, Kansas State Univ).

Note: A photon cannot literally be absorbed into an electron. This is a figure of speech to explain how an electron temporarily absorbs the energy of a photon.

Coming back to incident light, it was just mentioned that such light, once absorbed and emitted, can have a journey-time that can vary depending on the physical distance between each moving photon in a light ray. Light that is absorbed/emitted is referred to as 'incident light' or 'refracted light'.

The type of material or medium receiving the light very much influences the time taken for the photons to be absorbed and then emitted as new incident light. This is why rays of incident light vary tremendously, one from one to another, in their mentioned journey-times.

The mentioned physical distance between any two moving photons is referred to as the wavelength of light. This wavelength (i.e. distance) is what tells the eyes and brain to see the colour that we are looking at.

There are many millions of different journey-times of incident light rays, and each different journey-time determines the colours we see and the overall energy of a given light ray. As the photons enter the eyes, the physical distances between the photons trigger the eyes and brain to see specific colours:

"It has to do with the special parts of the eye called rods and cones. These are what make the eye act much like a spectroscope when measuring absorption and transmittance of light into and out of a substance" (source: K. Sundeen, Spectroscopy, University of Pennsylvania MCEP).

So just about everything we see (trees, streets, people, books, food, etc) is not direct light or reflected light, it is incident light that occurs as a result of absorption and emission, into-and-out-of the objects we see. This incident light arrives at our eyes as light rays with different journey-times. Each light ray will have its own mix of distances between moving photons (i.e. own mix of wavelengths), mapping out the panorama of colours and shapes that we see. This explains why light never bounces or reflects off anything.

To avoid confusion in terminology the following image shows some of the terms used in contemporary physics regarding light:

Column A	Column B
Incident light	Non-incident light
Coherent light	Incoherent light
Polarised light	Unpolarised light
Attenuated light	Non-attenuated light
Refracted light	Non-refracted light
Monochromatic light	Polychromatic light

Copyright © *Final Theory of Light: & Finding Extraterrestrials*

In the above image the phrases in column A are interchangeable and they all mean **exactly** the same thing. They all refer to the same process by which photons are absorbed and emitted from electrons inside atoms.

Equally, the six phrases in column B mean **exactly** the same thing. They refer to light that has been created, for example in a candle or the sun, but has not yet been absorbed and emitted from the atoms of some object, medium or material, i.e. it is disorganised light that carries a mix of different wavelengths.

The many different phrases referring to exactly the same kind of light have gradually arisen as a result of a poor understanding of the nature of light, and also because of the '*Big Misunderstanding of Light*' as explained in this book.

The phrase 'white light' causes endless confusion, so here is a clarification. The colour white, such as white paint or a white sheet refers to a colour that looks white. It looks white because when you have an equal mix of red, green and blue, the result is white. For example, if you mix red, green and blue lights for illuminating a football stadium you will have so-called white light, giving a good approximation to daylight. This kind of white light is incident light with a fixed combination of wavelengths (a fixed 'recipe') giving the colour white.

But sometimes non-incident light can also look white. For example, sunlight shining through the clouds can look white. Or some kinds of laser light or torchlight can look white, yet such light is incoherent, it is non-incident light, hence the confusion. More about this later in the book.

To finish on the subject of light attenuation it should be mentioned that the rate of attenuation varies tremendously. The 'rate of attenuation' refers to the percentage of light absorbed that is successfully emitted out as incident light.

For example, a pair of shoes may have a 52% rate of attenuation, meaning that for every 100 photons absorbed into the shoes, only 52 photons are emitted as incident light. The other 48 photons absorbed into the shoes were changed into heat or into other types of particles. A very good quality mirror may have a 99.9% rate of attenuation, meaning that nearly all the photons going into a mirror were emitted as incident light. Lead has nearly a 0% rate of attenuation, meaning that when you shine light onto lead, virtually all the photons that go into lead are not re-born as incident light (i.e. virtually no incident light comes out of lead).

Regarding planets, it's a similar situation. The moon has an 11% rate of attenuation, meaning that only about 11% of the sunlight absorbed into the moon is 'reflected back out' (attenuated) as incident light. For the Earth it's about 30%, for Mars 25% and so on.

The scientific name given to the mentioned rate of attenuation is the 'Bond albedo' effect. Here is a chart showing the Bond albedo effect for various planets in our solar system:

Name	Bond albedo
Mercury[2][3]	0.088
Venus[4][3]	0.76
Earth[5][3]	0.306
Moon[6]	0.11
Mars [7][3]	0.25
Jupiter[8][3]	0.503
Saturn[9][3]	0.342
Enceladus[10][11]	0.81
Uranus[12][3]	0.300
Neptune[13][3]	0.290
Pluto[14]	0.41
Charon[15]	0.29
Haumea[14]	0.33
Makemake[14]	0.74
Eris[14]	0.99

Source: Bond albedo, Wikipedia.org

For example, in the above chart, the Bond albedo effect for Earth is 0.306. This means only 30.6 % of sunlight absorbed into Earth is attenuated and sent out into space as incident light. The other approximate 70% of this sunlight is lost to heating the surface of Earth. The planet Eris (about the size of the moon) wins the day with a 0.99 Bond albedo effect, meaning that it attenuates nearly all the sunlight received by virtue of having a mirror-like surface.

*

What is the journey-time of light?

All photons always move at speed c, whether they are incident or non-incident photons. This begs the question: *how is it possible for the journey-time of incident light to be slower than the journey-time of non-incident light?*

Here is a 'two string analogy' to explain how the journey-time of incident light can take longer than the journey-time of white non-incident (incoherent) light.

TWO STRING ANALOGY

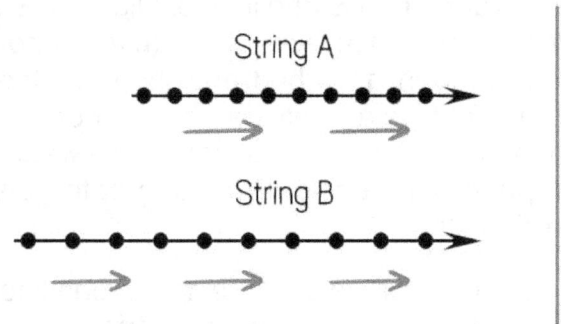

We imagine two pieces of string, string A is 1 metre long and string B is 1.5 metres long, as shown in above image. In string A we make 10 equidistant knots. In string B we do the same. (This has nothing to do with so-called 'string theory'). Each string represents a light ray, a stream of photons. String A is non-incident. String B is incident light.

So we imagine that all 20 string-knots (the total of both strings) represent 20 photons moving, say, from left to right in the analogy, at the same speed of light c. We also imagine there is a vertical line painted across the road as in the above image, and that both strings are moving towards the line, completely abreast of each other.

As each string crosses the line they continue onwards at the same constant speed of light. The leading knots in strings A and B continue to stay completely abreast because they are both moving at the same speed c.

But consider string B (the incident light). It is longer than string A because every knot (every photon) in string B has been absorbed and then emitted, thus causing a slight delay between each knot (between each photon). Thus, the distance between each moving photon in string B is a greater distance compared to string A.

As both strings cross over the same road line at the same time, the lead knots in each string continue to stay completely abreast of each other. A bystander uses an accurate timer and it shows the following:

String A took 3 seconds for the whole length of the string to cross the line. String B took 4 seconds for the whole length of the string to cross the same line. String B took longer because it was a longer piece of string by virtue of the greater distance between each knot. Remember that in each string the 10 knots are equidistant.

To summarise: in both strings we have a total of 20 knots (20 photons) all travelling at exactly the same speed c of light, never slowing down. Nevertheless, the whole journey-time of string B (with 10 photons) took longer than the whole journey-time of string A (also with 10 photons) to completely cross the road line. So although all photons (knots) in both strings travelled at the same speed c at all times, we can say that the overall journey-time of string B was a longer journey-time compared to string A. By taking a longer journey-time, we say that string B travelled 'more slowly'.

The **'Two String Analogy'** helps to explain how light can travel 'more slowly' (i.e. have a longer journey-time) even though every photon in the light ray travels at the same speed c at all times. In fact, virtually all the light that we see and receive on Earth is incident light. That is, light that is coming to us at a slower journey-time compared to the journey-time of white incoherent light.

The difference between the 'speed of light' and the 'journey-time of light' is that 'speed' refers to the speed of individual photons (i.e. speed 'c'). Whereas, the 'journey-time' refers to the time it takes a whole given bunch of photons to travel from A to B.

Now that we know the meaning of 'incident light' and 'journey-time of light' here is a question for you, dear reader: *when light travels through water, will such light regain its prior shorter journey-time upon leaving the water?* Are you ready for the answer ....? The answer is that when the light leaves the water it stays at the same longer journey-time indefinitely because it has become incident light as a result of going into water.

WHEN LIGHT LEAVES A MEDIUM SUCH AS WATER ITS SPEED AND JOURNEY-TIME STAY EXACTLY THE SAME AS WHEN IN THE MEDIUM

The speed c of light does not change at any point, either before, during or after leaving a medium. The photons that make up light can never change their speed. But the journey-time of light does indeed change as a result of going into a medium. As already explained, the attenuation (absorption/emission) of photons while in the water puts a tiny time-interval between the moving photons.

This attenuation stays fixed in the stream of photons even after leaving the water and will never change unless/until the photons are destroyed. If this stream of photons is examined in a spectroscope it will usually have a colour spectrum of blue. Thus, the time-intervals of the photons leaving the water will have a wavelength and a frequency of colour blue, which is a short wavelength. Why short? Because water molecules are packed very close together.

Hence, when photons travel from molecule to molecule to be absorbed and then emitted by each water molecule, the emitted photon travels a very short distance to the next molecule (a distance of empty space inside the water of about 0.31 nm). As a result, a very short attenuation time-interval is put between the stream of photons travelling through the water - this explains why we see the colour blue in water.

Note: when the photons in water travel through the mentioned empty space inside water, from molecule to molecule, they travel at the full speed 'c' of light. This is why we can say that the speed of light (but not the journey-time of light) does not slow down or change when moving through a medium.

"The speed of light never changes in, for example, water. The light stops for a moment at each atom and pays a short visit. In between atoms the light moves at the usual speed of light" (source: How Does Light Slow Down, space.com, July 2023).

There is a very widespread misconception regarding seeing the colour blue in water. Here is an erroneous example: *"Water blueness comes from the water molecules absorbing the red end of the spectrum of visible light. To be even more detailed, the absorption of light in water is due to the way the atoms vibrate and absorb different wavelengths of light".*

Every part of the above quote is incorrect. Water molecules do not absorb light at all, but the electrons in the atoms of water-molecules do indeed absorb light. Furthermore, when such electrons absorb light they do not absorb 'the red end of spectrum' - they absorb the whole photon and then emit a 'replacement' photon (this phenomenon of attenuation is well known to science). Last, but not least, molecules may or may not vibrate for all sorts of reasons, but such vibration does not make molecules absorb 'different wavelengths of light'.

This and similar statements in academic circles reflect the very widespread misunderstanding about the fundamental nature of light.

*

What is the frequency of light?

This is the most important question (and the most misunderstood question) when it comes to understanding the true fundamental nature of light. Once you understand the nature of the frequency of light, just about everything else regarding light falls into place. So what follows is a description of the frequency of light.

Note: It is important for your good understanding that you do not come to this page cold. That you have read the previous pages of the book up to this point.

We have said that in general, incident light has a longer journey-time than non-incident light. Also, every ray of incident light has a different time-interval between its moving photons compared to other rays of light. This time-interval determines the distance between the moving photons, and this distance is referred to as the 'wavelength' in optical physics. The length of the mentioned wavelength determines the number of photons in a given light ray. And this number is the so-called 'frequency' of a light ray.

In other words, the time-interval between moving photons determines the light's frequency. Longer time-intervals equate to a longer frequency. Shorter time-intervals equate to a shorter frequency. Think of the word *'frequency'* as a shorthand way of referring to the *'time-interval between each moving photon in a ray of incident light'*. It is more convenient to just say 'frequency'. So the frequency of light is nothing more than the measurement of the time-interval between the moving photons because this time-interval determines the concentration of photons in a light ray. The degree of concentration is referred to as the frequency.

How does the standard model of physics describe the frequency of light? To answer this let us consider the following image:

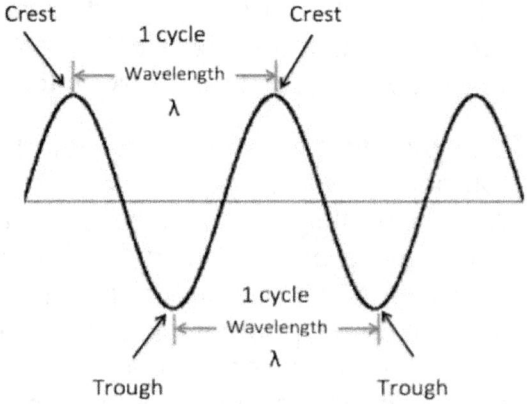

Crest, Trough and Wavelength

In the above image the wavy line is how the standard model of physics likes to depict the frequency of light. Such depiction is entirely artificial - it is a human construct created in a spectroscope to help scientists interpret the results of light analysis. One complete electromagnetic **wave cycle** is the *movement* of a wave from one wave crest to the next wave crest. An electromagnetic **wavelength** is the measurement of the *distance* from one wave crest to the next wave crest (i.e. from one photon to the next photon). So the wavelength represents the distance between moving photons. Electromagnetic **frequency** refers to the amount of wave crests (i.e. number of photons) that roll by in spectroscopy equipment every one second.

It's worth reading the aforementioned paragraph a second time - it will give you a greater understanding of light than many students of optic physics.

So the terms frequency and wavelength are somewhat interchangeable. The longer the time-interval between moving photons, the longer the wavelength, making the wavy pattern flatten out. The shorter the time-interval between photons, the shorter the wavelength, making the wavy pattern bunch together more.

You will appreciate from the aforementioned that a single photon cannot have a wavelength, and that the wavelength refers to the space (real physical distance) between moving photons, not to the photons themselves. And this space is determined by the time-intervals between moving photons.

"Photons are discrete particles that have a certain amount of energy, but not a wavelength because they are not waves" (source: Gregory A. Davis, Light, Wave or Particle? Fermilab, US Dept of Energy).

The extent of the time-interval between moving photons is determined by the time it takes photons to be absorbed/emitted (the attenuation) - this varies widely, depending on the type of medium or material emitting the light before it reaches a spectroscope. Also, the extent of time-intervals can be artificially induced - for example radio equipment induces long time-intervals (long wavelengths), and X-ray equipment induces short time-intervals (short wavelengths).

We need to keep in mind that the frequency of light is entirely dependent on the degree of slowness in its journey-time. In a spectroscope the slower the 'roll by' journey-time of a light ray, the lower its frequency. And the quicker the 'roll by' journey-time of a light ray the higher its frequency.

The main purpose of a spectroscope is to work out the frequency of light. It is designed to analyse the received incident light and work out the density or concentration of photons in the light received. This is done in a roundabout way since no spectroscope would be capable of counting the number of photons in light while many millions of photons go through the spectroscope at the speed of light.

When a spectroscope receives incident light it takes a sampling of such light to establish the frequency. As the light rolls by a fixed point in a spectroscope it 'captures' the amount of light that has rolled by in one second of time. This one-second sampling is then channelled through a prism or a grating inside the spectroscope to establish the colour spectrum of said light. And the colour spectrum tells you the frequency. The colour spectrum of light is entirely determined by the density of photons in light. And the density is entirely determined by the time-intervals between each travelling photon.

So the time-interval between moving photons is what determines the frequency, and the time-interval also corresponds to a particular colour spectrum. Ergo the colour spectrum of a particular bunch of photons reveals its frequency. Each frequency has its own associated colour as shown in the following image:

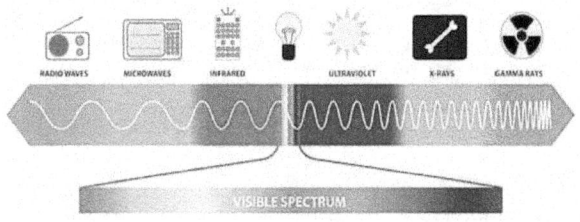

This image shows the different frequencies of light, referred to as the electromagnetic spectrum. For example the wavelengths of X-rays or gamma rays are very short, i.e. they have very small time-intervals between their moving photons and therefore very small distances from wave crest to wave crest. As a result the waves are much more bunched up compared to the longer waves of, for example, radio waves.

*

What is the amplitude of light?

The so-called 'amplitude' of light does not exist as something real, it is entirely a mathematical concept. As explained in the previous section, the movement of light is often depicted as a wavy line with crests and troughs. But the image of a wavy line is simply a diagrammatic way of explaining certain aspects of light. It does not represent how light moves in reality.

Spectroscopes are designed to analyse certain characteristics of light and give the results as digital numbers. But to help scientists understand such results, spectroscopes can also translate the digital numbers into an artificially created wavy line that can be seen through a viewing port on the spectroscope or on a computer screen. Here is an image of this mentioned wavy line showing the amplitude of light:

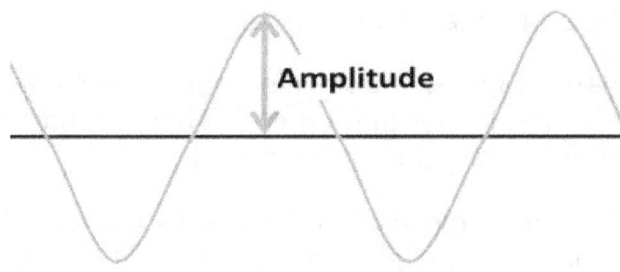

In the above image the amplitude is shown as being half the total height of a wave from crest to trough. The bigger the amplitude, the more bunched-up and taller the waves. The shorter the amplitude, the less bunched up and shorter the waves. Imagine that you hold each end of a wavy line and you pull the whole wavy line wider: this will flatten the wavy lines and reduce the amplitude.

The concept of the amplitude of light has arisen as a convenient way of referring to the extent of the bunched-up waves. Instead of saying *'the waves are very bunched up'*, it may be more convenient to say *'the amplitude is X amount'*. The numerical value of X gives a measurement as to how much the waves are bunched up.

The amplitude indirectly tells you the distance from crest to crest. This tells you how bunched-up the waves are. And this in turn tells you the frequency, i.e. the density or number of photons in a given light ray. Remember that the distance from one crest to another is the distance *between* each moving photon in a light ray. So the shorter this distance, the more bunched-up the moving photons.

Hence, for example, the more bunched-up the waves, the longer the amplitude and the greater the energy and brightness of the light.

Thus the amplitude of light does not refer to any kind of real physical property of light. It simply refers to a mathematical way of calculating the frequency of light based on a fictitious wavy line that is artificially created by a spectroscope.

*

How does light move?

Although scientists cannot directly see how light actually moves, the consensus is that light moves as a sinusoidal vibration - a kind of vibrating pattern based on the electromagnetism of photons. This was first proposed by James Maxwell (1831-1879) by understanding the sine and cosine movement of electricity and magnetism that act together to propel light forward. Here is a photo of Maxwell in front of a painting backdrop:

The following image shows the oscillating interaction between the electric and magnetic fields that make up a photon of light.

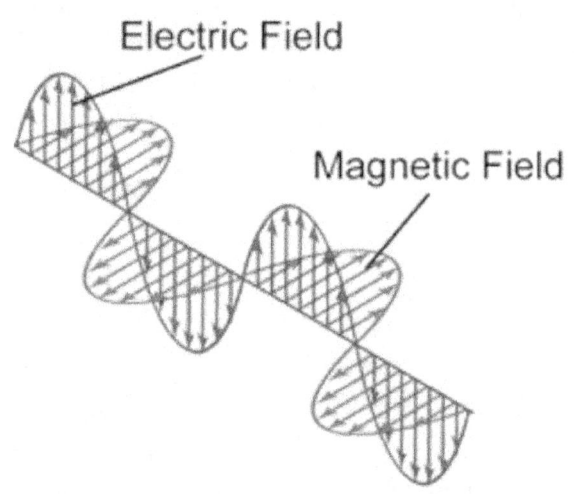

As already discussed, a photon is a little packet of electromagnetic energy (EM energy). It is well known in physics that EM energy is produced from a change in the electric field that causes a change in the magnetic field just ahead of it. This then causes a change in the electric field just ahead of that, and so on and so on. This self-propagating change in the electric and magnetic fields is referred to as an oscillation (or a vibration) between the electric and magnetic fields of energy, thus propelling the photon forward at the speed of light 'c'. So put simply, electromagnetic oscillation propels light forward.

As the electricity and magnetism of a photon oscillate, they renew each other and they can keep going indefinitely at the speed of light. The amazing thing is that no energy is lost; the photon is completely self-propagating.

The sinusoidal movement of photons is more akin to a vibration than to a wavy pattern with crests and troughs. And as the photons move forward with a sinusoidal vibration, they nevertheless always move in straight lines, radiating out in all directions as in the following image:

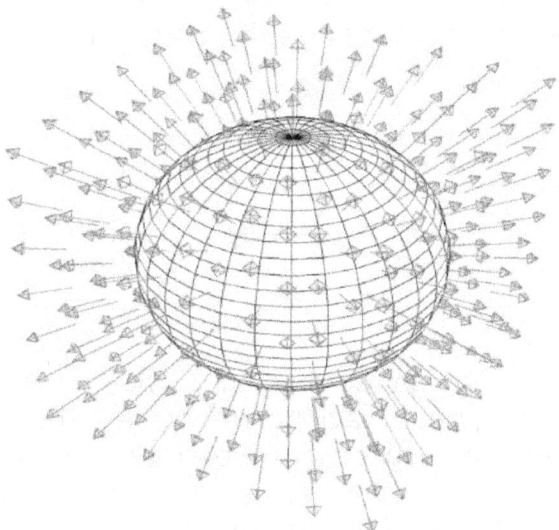

If the light comes from a focused source such as a light gun, a torch, or a halogen lamp, the light will expand in all directions, but more akin to the growing face of a cone.

Samples of light can be captured in spectroscopes for detailed analysis. As the light goes past a fixed point in a spectroscope, a one-second sample is taken as the light rolls by. As already mentioned, this one second sampling is then channelled through a prism or grating inside the spectroscope to establish the colour spectrum and other information.

The spectroscope results arising from a one-second sampling can then be displayed in various formats. Typically, scientists study the so-called 'spectral lines' in such results because they can reveal much information such as chemical compositions, signs of biological life, types of elements, and so on.

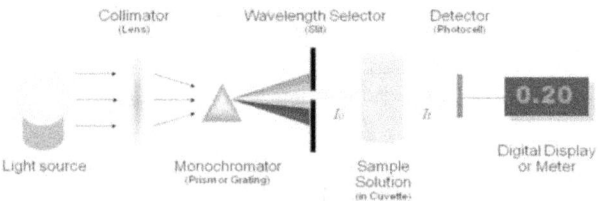

This image shows a digital display (to the right in the image), but a wavy pattern can also be artificially created in a spectroscope viewing port (or computer screen) to help scientists understand what they are looking at. The wavy pattern with crests and troughs seen in spectroscopes is entirely a human construct - it does not show any kind of movement of the light itself, nor how the light received has actually moved.

When we see light we are looking at many millions of photons moving in straight lines in all directions. When some of those straight lines meet our eyes, that is the light that we actually see.

When light moves it cannot slow down or speed up or stop. Also, light can never bend and it can never bounce or reflect off anything. It can only ever keep moving in a straight line at the given speed of light (about 300 million metres per second, denoted as 'c' in physics).

*

Why is the constant speed of light constant?

The speed of light at 'c' is constant because of the **law of energy conservation**. All light is made of moving photons and nothing else at all. When light is first created from a light-source (light bulb, candle, sun, etc) it means that heat in the light source has over-energised the electrons inside the atoms of the light-source.

All electrons in the Universe are identical to each other, regardless of the type of atom housing the electrons. Hence the tolerance level (of being over-energised) is the same for all electrons. In other words, when electrons are over-energised they all absorb exactly the same amount of excess energy because electrons everywhere are identical.

These over-energised electrons become unstable and are compelled to get rid of their excess energy almost immediately by releasing such energy in the form of newly created photons. So all electrons, regardless of the type of atom, will absorb and then emit exactly the same amount of energy so as to go back to their previous state of energy. This ensures the overall amount of energy in an atom is maintained (the **law of energy conservation**). If it were not for this, atoms would become unstable, fall apart, and goodbye to all life in the Universe.

So when photons are created by electrons, it means that all photons are created with the same amount of energy, and indeed all photons in the

Universe are identical to each other in every way. The electron's surplus energy that is used in the creation of a photon gives the photon an initial kick of energy that sends the photon shooting out of the atom at the speed of light.

This initial 'kick of energy' is enough energy to kick start one electric oscillation which in turn generates one magnetic oscillation. So when the electric oscillation occurs it is depleted of all energy by transferring its energy to magnetism. This enables the magnetism to then oscillate and in so doing transfer its energy to electricity. So the two halves of the photon (electric and magnetic halves) become self-propagating - they renew each other by 'leapfrogging' over each other in what is known as a transverse sinusoidal movement. This transverse movement is what makes the photon move forward, but always in a straight line.

Given that all electrons are identical, it means that all photons are created equal, with exactly the same amount of oscillating energy. This ensures that all photons oscillate at the same speed, thus making all photons in the Universe move at the same constant speed of light 'c' (often referred to as the invariant speed of light).

Note: The electromagnetic oscillations of photons have everything to do with galvanising their movement forward, and have nothing to do with the energy of light. There is no relationship at all between the oscillations of light and the energy of light.

Physicists have puzzled over the constancy of light-speed for many years. Light-wave theory, based on false concepts about how light moves, is very entrenched in contemporary physics and as a result it has held back an understanding of how light behaves and why light speed is always constant. As revealed in these pages, we now know at last exactly why the speed of light is constant.

*

What is the energy of light?

The energy of a photon is given as the energy of a single oscillation of a photon. This does not mean that with each oscillation the photon gains more energy. Rather, it means that with each oscillation the photon has the same amount of energy, referred to as the 'energy constant' of light. It is calculated that the energy of any single photon is very, very little, barely 6.6 joules. Because of this, the usual way of referring to the energy of light is based on the density of photons (i.e. the number of photons) passing through a spectroscope in one second of time. This is explained more fully in what follows.

> **Every photon in the Universe is identical and therefore has the same amount of energy, namely 1 hertz (about 6.6 joules)**

Note: A joule is how we humans measure energy. Here's an analogy. A joule is like a measuring cup, and photons are like the liquid (energy) that fills the cup. The cup (joule) is not made of the liquid (photons), but it can be used to measure the amount of liquid (energy).

The fundamental constituent of heat itself is a somewhat controversial subject due to confusion in semantics. It is said that heat is not made of photons because heat refers to the movement of atoms within an object and we feel that movement as heat. But fundamentally, that movement of atoms over-excites electrons inside the atoms making the electrons emit photons. It is those photons that we feel as heat. So it is entirely correct to say that all kinds of heat are made of photons.

All types of light (from radio waves to gamma waves) travel at the same speed of light. The speed is determined by the universal rate at which the electricity and magnetism of a photon oscillate. So all photons oscillate at the same rate and hence move at the same speed of light c.

To clarify further, all 'types' of light are a form of radiation that moves outward as streams of photons. By 'types' this refers to sunlight, torchlight, X-rays, radio waves, etc. They all consist of the same kind of light and the same identical photons. The only difference between the various types of light is the time-interval between its moving photons, i.e. the frequency. Each type of light has its own measurement of frequency as already discussed.

Here's a question for you, dear reader: *Do all photons in the Universe have the same amount of energy?* While you think about it, here is the same question phrased differently: *Can single photons have different amounts of energy? Are you ready?* Here goes: To the first question the answer is YES, all photons have the same amount of energy. To the second question the answer is NO, single photons cannot have different amounts of energy.

"In a light beam of a certain power, a corresponding amount of energy will be carried (as photons) by the beam in any given time, and this is related to the photon density" (source: Photon Density, Fibercore.humanetics.com).

Put simply, the energy of light relates to the density (amount) of photons in a given light ray. Since every photon carries the same amount of energy, the energy of light is not determined by how much energy is carried by each photon, it is determined by the number (the density or concentration) of photons in a light ray.

Many physics textbooks and research papers state that photons can have different amounts of energy. This topic is full of confusion and misunderstandings. Why so? Because the terms 'wave cycle', 'wavelength' and 'frequency' are usually misunderstood. And indeed the propagation of light itself has not been fully understood.

Quite simply, the energy of light is determined by the concentration of photons in such light. In other words, the density of photons determines the amount of energy in different types of light radiation. And the density of photons is defined as the frequency (think: how frequent) of photons that roll by in one second in a spectroscope or wavemetre. For example, X-rays have much more energy than daylight rays because X-rays have a much higher density of photons (a higher frequency) compared to daylight rays.

Clearly there is only one type of light, so when reference is made to 'different types of light' this should read 'different light frequencies'. And remember that the frequency of light is derived from the time-interval between the moving photons in a light ray.

Just to be clear, one X-ray photon has the same energy as one daylight photon or one radio wave photon. But X-rays have a much higher density (i.e. frequency) of photons compared to, say, daylight. The high density (concentration) of photons in X-rays is achieved by using very high voltage equipment that shoots out a high number of photons in a very short period of time, like shooting photons out of a machine gun; this is what gives X-rays very high energy.

If you were to examine a single photon without knowing its source it would look exactly the same as any other photon. In other words, all photons are the same everywhere.

Also, there is only one type of energy in the Universe. In physics energy is defined as the force that makes things move. So in this sense there is only one kind of energy, and at a fundamental level such energy is based on photons. For example, electricity is a flow of electrons along a wire. The electrons are made to move along a wire by voltage energy. And voltage energy arises from photodiodes which in turn arise from the movement of photons.

Whatever the kind of energy in physics, if you drill down far enough you discover that photons are at the root of the energy in one way or another.

*

What is the big misunderstanding of light?

As discussed, the electromagnetic oscillations of photons make light move forward at the speed of light. These EM oscillations galvanise photons forward in a vibrating sinusoidal movement, albeit always moving in straight lines. Each oscillation is a vibration. But it does not follow that the EM oscillations of photons determine the energy or frequency of light, hence the big misunderstanding.

And it does not follow that light moves as so-called 'light-waves', i.e. as single fields of energy containing multiple photons. The wave theory of light (born from this big misunderstanding) postulates that all the photons in a light-wave oscillate at the same rate, and that any disturbance to a part of the light-wave affects the whole light-wave.

Light-wave theory has great difficulty explaining the frequency of light when it is based on the rate of EM oscillations. This is so because it is increasingly being realised in contemporary physics that EM oscillations have nothing to do with the energy or frequency of light.

In short, a big misunderstanding arises by making a false association between EM oscillations and the frequency of light. Unfortunately, this misunderstanding has led to other mistaken beliefs such as a belief in a particle/wave duality of light, and a mistaken concept of the frequency of light. Once this big misunderstanding is seen for what it is and dismissed, it greatly helps to understand the true nature of light.

At the heart of the big misunderstanding is to postulate that the distance travelled by light during one complete electromagnetic oscillation is equal to one wavelength. This fatal flaw in understanding the nature of light gave rise to the big misunderstanding described in these pages. EM (electromagnetic) oscillations have nothing to do with wavelengths, and hence nothing to do with light frequency.

You will know from reading this book that the wavelength is the real measurable physical distance between any two moving photons. The description of the frequency in physics is based on the number of such wavelengths that occur in one second of time as the light goes through a spectroscope.

So the frequency is the concentration of wavelengths which is the same as saying the concentration of photons. The smaller the wavelengths (the distances between photons) the greater the concentration of photons in a given light ray. The length of a wavelength is entirely determined by the speed at which electrons are able to release (create) photons one after another. The electromagnetic oscillations of photons have everything to do with the movement of light and nothing to do with the wavelength or frequency of light.

The energy variation of light comes from the number of photons concentrated in a light ray, given that the energy of each photon is the same. If the time-intervals between separate moving photons (separate moving packets of energy) is a small time-interval then logically the concentration of photons in a light ray will be more numerous and the energy of the light ray will be high. The frequency is determined by the wavelength, i.e. the distance or length between two moving photons in a given sampling of a light ray. Photons are the same everywhere, so any individual photon will have the same energy as any other photon.

*

Insurmountable contradiction

An insurmountable contradiction arises in light-wave theory when it is postulated that the frequency of light is determined by its rate (speed) of electromagnetic oscillations. And that the rate of oscillations also determines the energy of light. It is well-established in contemporary physics that an electron can only release a photon of the same energy as the one it absorbed. So regardless of the type of material or medium absorbing the photon, the incident photon emitted will have the same energy as the photon that was absorbed. This ensures that all photons have the same rate of EM oscillations and hence move at the same speed of light everywhere. In effect, the universal constant speed of light is set by electrons.

The insurmountable contradiction facing light-wave theory is the following. Since electrons always release photons of the same energy as those absorbed, this means that the speed of electromagnetic oscillations does not change. And given this, if the rate of oscillations does not change, then how can light-wave theory postulate that different rates of oscillation determine different rates of light energy!

Let us remember that light-wave theory clearly says that the rate of electromagnetic oscillations determines the frequency. That is, the rate of oscillations determines the colour that is seen.

But it is well-established scientifically that the rate of oscillations of a photon does not change when it is attenuated. So if the rate of oscillations determines the colour we see, how does wave theory explain that the colour red is seen when looking at a red car? That is the contradiction.

That insurmountable contradiction will remain for as long as energy and frequency (and wavelength) are believed to be determined by the rate of electromagnetic oscillations.

Note: Individual electrons can temporarily have different amounts of energy, but not photons. Also free electrons can have more energy. A material with high-energy electrons such as crystals and certain non-metal-chemicals will absorb and emit photons more quickly. This reduces the time-intervals between the emitted photons, and this in turn increases the energy and frequency of a given stream of photons.

When it is said that electrons always release photons of the same energy as the ones absorbed, this can cause confusion because electrons can artificially be made to emit very high-energy photons (higher than the photons received or absorbed). For example lasers and X-ray machines can do this. The confusion arises over the phrase 'high energy photons'. This phrase does not mean that each photon in a stream is endowed with high energy (a common misconception). It means that the streams of photons from, say, a laser or X-ray machine have a higher concentration of photons (the photons are more numerous and bunched up).

The following topics are examined briefly to explain further the big misunderstanding which is so entrenched in contemporary physics:

1. The brief history giving rise to the misunderstanding.

2. Double slit experiments.

3. Contemporary wave-theory of light.

*

The brief history giving rise to the misunderstanding

In 1665, Italian physicist Francesco Mario Grimaldi (1618 to 1663) discovered the phenomenon of light diffraction and pointed out that it resembles the behaviour of waves. By not understanding at the time the full nature of light refraction and diffraction, the belief in light-wave theory was born. A few years later Christian Huygens believed that light was made of waves propagating perpendicular to the direction of its movement. It is now known that EM oscillations do indeed function in a perpendicular motion to each other thus propelling light forward in a vibrating sinusoidal movement. But it doesn't follow that EM oscillations determine the energy and frequency of light.

This misunderstanding was further compounded by Christiaan Huygens in 1678 when he said that light is a longitudinal wave that sets light particles oscillating in phase with each other in the form of a wavefront. Huygens based his wave theory on the existence of an all-pervading mysterious ether through which the light-waves travelled.

Then in 1803 Thomas Young started the so-called 'double slit experiments' showing (albeit falsely) that light moved as a wave rather than as streams of separate photons. In this scenario, light frequency was explained by the speed of oscillations of a light-wave.

A hundred years later the classical wave theory of light was firmly established in physics when Albert Einstein pronounced in 1905 that light possessed a duality of particle and wave characteristics. This gave much credence to the wave theory of light from then onwards. It is quite possible that Einstein (among others) was 'taken in' by the double slit experiments of the day.

The confusion in Einstein's mind is shown by the fact he first hypothesised that light is made of particles. But he later changed his mind saying that light is also a wave. He then linked the intensity of light to the number of photons in a beam, which is correct. But then later stipulated that the frequency of light is determined by how much energy each photon carries, which is incorrect.

When scientists apparently see 'wave characteristics' of light, any such characteristics can readily be explained away by understanding the true fundamental nature of light as explained in this book.

*

Double slit experiments

The belief in the veracity of light-wave theory has been greatly boosted by so-called double slit experiments. Such experiments consist of shining a beam of light towards a metal sheet with two slits, as depicted in the following image:

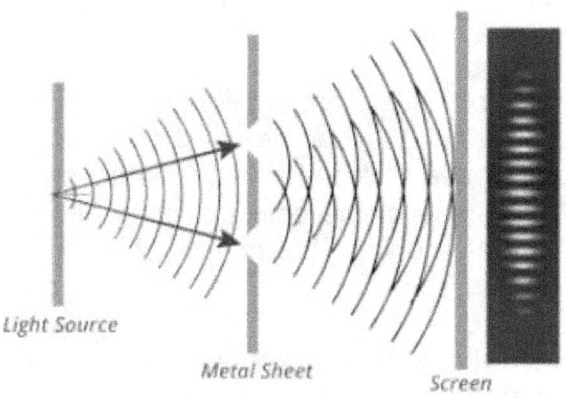

The screen shown to the right in the image is a photon detector and in this image it is seen to detect more photons in the middle rather than at the sides. The conundrum is that when light goes through the two slits aligned to the sides, the photons nevertheless hit the middle of the detector screen more often. How can this be?

Wave theory advocates are quick to say that light goes through each of the two slits as a wave, and then having gone through the slits, the light reforms into waves again, with each wave expanding so as to give two expanding waves going towards the screen. In doing so it is claimed that each wave overlaps (that the waves interfere with each other), thus sending more photons to the middle of the photon-detector screen. It is argued that this shows that light must be travelling as waves because if the light had been travelling as just streams of photons in straight lines (albeit in a sinusoidal vibrating movement) they would not be hitting the middle of the screen with a greater preponderance.

Is this so? Given that the light source (to the left in the above image) shoots a stream of expanding photons, these photons move forward as a growing cone. The face of the cone reaches all parts of the metal sheet shown in the middle of the image. Then the photons start to go through the two slits. At this point the scenario that we have is equivalent to two torches, with each torch shining through each slit (as if there are two separate sources of light).

So each slit becomes a separate source of light, and each slit creates another growing cone of light going towards the detector screen. As shown in the above image, the spreading cones overlap and this results in a greater preponderance of photons hitting the middle of the screen.

There are two factors to consider in explaining the misunderstanding of double-slit experiments.

Factor one: refraction and diffraction. As the photons travel from the light source to the photon detector screen they will be refracting with air and with the surrounding partitions (walls, floor and ceiling) of the double-slit apparatus. This will be happening both before and after the light goes through the two slits.

In particular, air contains mostly oxygen and nitrogen, and small amounts of other gases such as carbon dioxide, neon, hydrogen and water vapour. All these constituents of air serve to absorb and then emit new photons in all directions. This in turn helps to spread the streams of photons in all directions as they travel through the apparatus.

Regarding diffraction, when the photons go through each of the slits, many photons will diffract as they go through the slits. Here is an image to explain this:

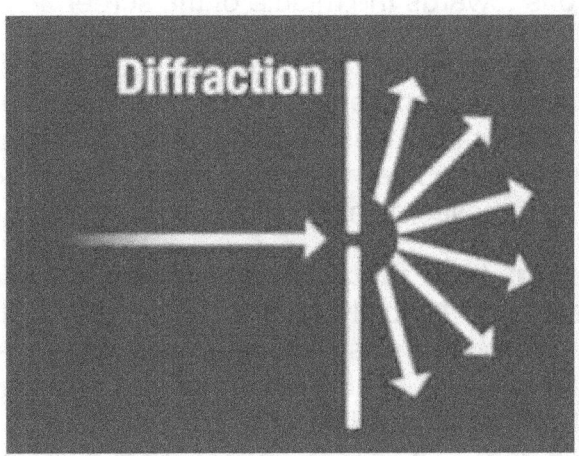

The difference with diffraction (compared to refraction) is that when the photons are absorbed into the edge (the lip) of the slit they are going through, they are emitted as new photons and sent on their way at 'all angles' (see the following note), not just at one particular angle as when light hits air or water for example.

Note: Diffraction occurs as a result of Snell's Law. This explains how and why the photons emitted do so at certain angles. More about Snell's Law later in the book.

Hence, when the photons go through the two slits they are not going through at a particular angle of propagation. They go through each slit and then become two new expanding cones of light moving forward towards the detector screen. As they do so the two growing cones overlap, sending a majority of photons towards the middle of the screen.

It is quite likely that in the 19th Century heyday of double-slit experiments the full science of refraction and diffraction was not fully appreciated, thus misleading double-slit experimenters into believing that light moves as single fields of multiple photons.

Factor two: metal sheet emission. Double slit experiments typically involve using a thin sheet of metal for the two slits. The belief was that metal would not absorb and emit new photons to any significant level so as not to disrupt the experiment. It is now known that this is not so because a metal sheet (unless made of lead or uranium) can refract up to 70% of the photons received. And when this happens, the refracted light would travel to all sides of the boxed area between the light source and metal sheet in the middle (undergoing further refraction), greatly contributing to the mix of photons going in all directions.

By not understanding these two factors, many people have been led to believe that double slit experiments show that light moves as waves and has the characteristics of both light-waves and light particles. This in turn has served to consolidate the mentioned big misunderstanding.

*

Contemporary wave-theory of light

Let us remember that light is not made of streams of particles, it is made of streams of packets of energy. Contemporary wave theory says that these streams of photons (packets of energy) travel together as a single field of energy containing many photons, all oscillating at the same rate. This single field of energy is referred to as a 'light-wave'. Furthermore, as mentioned it is postulated erroneously that the energy of light arises from its electromagnetic oscillations.

The standard model of physics correctly says that light radiation can vary in strength, i.e. in its energy. As discussed, the electromagnetic (EM) spectrum depicts a wide range of many types of EM radiation, such as radio waves, microwaves, infrared light, ultraviolet light, X-rays and gamma-rays. All these types of radiation are made up of streams of photons, all of which are identical to each other. The only thing that sets apart the different types of EM radiation is the time-interval or distance between the moving photons.

A stream of photons in a radio wave will have longer time-intervals between its moving photons compared to the time-intervals in a stream of X-rays. As already explained, the time-intervals between moving photons are referred to as the wavelength from which the frequency of light is derived. So the frequency of light is determined by the density or concentration of photons in a given sampling of light.

But current wave theory does not agree with this. The frequency of light is regarded as being determined by the rate of EM oscillations in light-waves. It is considered that the rate (speed) of electromagnetic oscillations is set the moment light is created, and thereafter does not change until the light is absorbed or destroyed. Thus light-waves are said to be created with different rates of EM oscillation according to how the light was created.

As an example let's consider an X-ray machine and how it creates X-rays. As mentioned, an X-ray machine is designed to use high voltage electricity to produce a burst of photons in quick succession to give a high frequency. This high frequency (i.e. density) of photons is what gives X-rays high energy. An X-ray machine has no effect on the rate of electromagnetic oscillations of the photons.

"An X-ray is a packet of electromagnetic energy (photons) that originates from the electron cloud of an atom. This is generally caused by energy changes in an electron, which moves from a higher energy level to a lower one, causing the excess energy to be released as photons" (source: X-rays, Australian Radiation Protection and Nuclear Safety Agency).

"All electromagnetic radiation consists of photons, which are individual packets of energy. The rate at which energy is delivered by a beam is determined by the number of photons in each ray of energy" (source: Radiation Safety Training for Analytical X-Ray, Western Kentucky University, USA).

So the high energy of X-rays arises from the high concentration of photons in the rays. This high concentration of photons is referred to as a high frequency. But wave theory postulates that the frequency of light arises from the rate of electromagnetic oscillations of the photons. This contradicts the way that X-rays are produced, or for that matter, the way any other kind of light radiation is produced.

When confronted with this contradiction, some light-wave theorists say that when X-ray machines produce X-rays, they produce photons with high oscillating EM waves, and this in turn produces fields of coupled photons with high energy. That the high energy of X-rays is due to a wave of joined-up photons all oscillating at the same extra-high rate. In physics there is no evidence of 'joined-up' photons.

There is no escaping the mentioned insurmountable contradiction faced by light-wave theory that states emphatically: *"The frequency of a light-wave is determined by its source, so it remains the same when the light-wave travels through different mediums or when emitted from different objects"*.

But if the unchanging frequency of a light-wave (i.e. its unchanging rate of oscillations) is what defines the wavelength of light, then how can it also be said that such wavelengths determine the colours we see! That is the contradiction.

When it comes to refraction (attenuation), this subject is all about absorption and emission of photons as they go into and out of a medium or object. But light-wave theory has a hard time explaining how refraction happens, and there is a lot of confusion.

Why so? Because for refraction to occur, the wavelength of the newly emitted photons must have changed by becoming longer or shorter. The length of the wavelength determines the frequency of light, and the frequency determines the colour spectrum of such light.

However, according to light-wave theory, refraction is not about absorption and then emission of new photons. It's about light undergoing a change in speed and wavelength as it goes through a medium. The speed of the light-wave is said by some to change while in the medium and to then revert to its prior speed when leaving the medium as depicted in the following image:

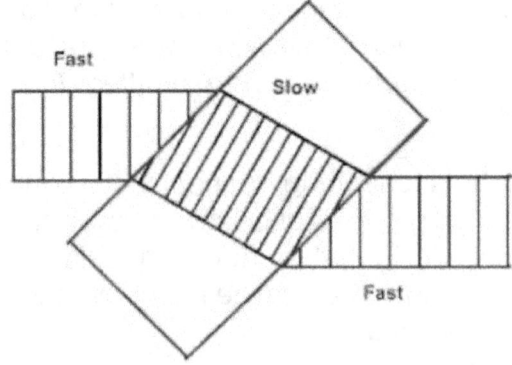

Regarding the wavelength, light-wave theory says that while in the medium the wavelength is changed by a change in the rate of EM oscillations caused by the medium. When the light comes out of the medium, it comes out with a changed wavelength and this is what conveys the colour of the medium it came out of. So in light-wave theory, refraction is all about a change in speed and a change in the rate of electromagnetic oscillations of the light.

But this explanation raises several questions and contradictions, such as the following:

* If the light-wave leaves the medium with a changed wavelength wouldn't this also foist a changed speed on the light-wave after leaving the medium?

* If refraction changes the rate of EM oscillations in the light-wave, do these oscillations stay changed once the light-wave leaves the medium? And if so, wouldn't this affect the speed of the light in this scenario, thus violating the constant speed of light?

* How does wave theory explain a change in speed of the light-wave while in the medium? What exactly is causing this change in speed?

* If a light-wave is a field of energy containing multiple photons (multiple little packets of energy) what happens to the photons during and after refraction? Do they all stay intact? If they change, how so?

* When light-wave theory says that refraction is the bending of light as it passes through a medium, what exactly causes the bending of light? If the answer is a change in speed, then what exactly causes this change in speed?

Here is an example of light-wave theory attempting to explain how a photon is absorbed and then emitted (the attenuation of light). The following is an abridged extract taken from a website dedicated to light-wave theory:

Quote

A photon may be absorbed by a particle, such as the electron, transferring energy from transverse ray form to longitudinal wave form. The interaction occurs not with the particle's standing waves, but instead it is the interaction with wave centres at the particle's core. The photon moves as aether granules towards an electron with wave centres at its core of the particle. A photon must match the right frequency to be absorbed by a particle because the interaction of the photon's components (granules) must match with the core of the electron (wave centres), and also it must match the resonance frequency of the particle.

Regarding 'stimulated emission' caused by photon absorption, an electron is first excited to a higher level. While excited, a second photon is used to excite the electron further. This results in two photons being generated that leave the atom. The two photons will be identical in energy, spin and polarisation. The electron vibrates and creates two photons travelling in opposite directions. Similar to spontaneous emission, one photon will leave the atom and one will reach and be absorbed by the nucleus as recoil.

Unquote

Quite apart from the incomprehensible rhetoric in the above explanation, the assertion that an electron will absorb one or two photons and then emit two photons is scientifically incorrect.

An electron can only absorb and then emit just one photon at a time. Once this process is completed the electron can again be ready to repeat the process if/when it is faced with absorbing another photon.

"An atom's electron can absorb or emit one photon when an electron makes a transition from one energy level to another" (source: emission and absorption processes, britannica.com).

"Is the following statement true?: a single photon excites only a single electron'. Yes, this is true. According to the photo-electric effect, each photon is able to excite only one electron across the band gap" (source: Toppr.com, an educational website with more than 6 million students).

"A single electron can emit only one photon at a time. When an electron transitions from a higher energy state to a lower energy state, it emits a single photon with energy corresponding to the energy difference between the two states. This is a

fundamental concept in quantum mechanics and the behaviour of light at the atomic scale" (source: ChatGPT, Jan. 2024).

When it is said that the speed of oscillations of a photon can be detected, thus proving light-wave theory, the reality is quite different. The oscillations of light itself aren't something humans can directly see, it is something we can measure and perceive indirectly through its effects. But those effects come from the rate of concentration of photons, not from the rate of oscillations of photons.

Finding straightforward comprehensible answers to questions about light wave theory bears little to no fruit so we will leave it there.

Last, but not least, let's consider several light-wave theory contradictions:

1. Speed of light contradiction. Light-wave theory is emphatic in saying two things: 1. Light-waves move forward due to the oscillating electric and magnetic fields. 2. The different levels of the energy of light are determined by the different speeds of the oscillating electric and magnetic fields.

The contradiction is simple: given that all light moves at the same constant speed, how can this be explained if the speed of oscillations can vary from light-wave to light-wave?

2. Light reflection contradiction. Light-wave theory is clear in saying that light-waves cannot be absorbed or go into certain solid objects. But light-waves can indeed go into water and other liquids. So at what point does a liquid become too thick or too solid for light-waves to enter? What is the cutoff point?

Also, quite apart from the fact that light cannot bounce or reflect off anything, light-wave theory clearly claims that we see a colour as follows:

"When light hits an object, the object reflects some of that light and absorbs the rest of it. Some objects reflect more of a certain wavelength of light than others. That's why you see a certain colour. For example, a lemon reflects mainly yellow light".

But if a light-wave or part of a light-wave is reflected off a lemon, how was the colour yellow acquired by the reflecting light-wave without going into the lemon? That is the contradiction.

3. Photoelectric contradiction. The so-called photoelectric effect is well-known and has been verified in many experiments. When light hits a solid material such as metal, some of the electrons in the solid material will be ejected completely from the material. Such ejection occurs immediately without any time delay because electrons tend to be loosely held by their atoms inside metal. This is why metals conduct electricity so well. Electricity is the flow of electrons.

The greater the frequency of the incoming light, the greater the concentration of photons in such light, and the greater the effect on ejecting electrons. So when high frequency light sends a barrage of photons into a metal, the photons can easily overpower electrons with too much energy causing them to be ejected instantly, without any attenuation occurring.

The photoelectric effect has never shown that electrons jump out with a time delay; any that do jump out always do so instantly. If ever there was proof that light-wave theory is spurious, this is it.

How does a light-wave theory advocate answer this? He/she may say that *"when a light-wave hits a solid material like metal, for example, the energy of the light-wave is uniformly spread over the wave front. As a result, there will not be enough photon-energy to immediately eject electrons. But eventually, as the photons in the light-wave build up inside the metal, there will be enough energy from such photons to expel some electrons after a short time interval"*.

There is a big problem with this answer - it has never happened. That is, however many or few electrons jump out, they always jump out without any time lag at all. Therefore light-wave theory has no answer to this contradiction.

Consider that electrons jump out by being overpowered with too much photonic energy. Such energy arises from the high concentration of photons hitting the electron. But a light-wave is said to generate its energy from the rate (speed) of its electromagnetic oscillations, not from any high concentration of photons.

When Einstein was confronted with this contradiction he felt he could not do an about face and deny light-wave theory altogether. He resolved the contradiction by saying that light sometimes behaves as streams of photons and sometimes as waves of photons. This led in part to Einstein receiving the Nobel Prize in 1921. It also very much helped to cement the big misunderstanding of light as explained in this book.

Light wave theory is very clear that light always travels as light waves, but in some circumstances it is manifested as streams of separate photons. This begs the question: What happens to light as it approaches a metal surface? Does the light mysteriously change to being streams of separate photons as it gets near the metal?

In contemporary physics, those who try to defend the photoelectric contradiction say the following:

"Light doesn't 'choose' to be a wave or a particle. Instead, we model it as a wave when we want to explain (or calculate) interference, but need to model it as a particle when we want to explain (or calculate) the photoelectric effect."

This is the same as saying: *"light does not choose whether to be a wave or a particle, instead we humans make that choice. We humans decide whether we want light to be a wave or a particle depending on the circumstances"*. This of course is very convenient, a copout, and is total nonsense.

Note: It is often (albeit mistakenly) said that Einstein was given the Nobel Prize in 1921 for discovering the photoelectric effect. In fact, the photoelectric effect was discovered in 1887 by the German physicist Heinrich Rudolf Hertz.

4. Blackbody contradiction. Light-wave theory cannot explain blackbody radiation. A blackbody is defined as an object that can absorb photons of light, but does not emit the same quantity of new incident photons to replace the absorbed photons. Blackbody is a generic term for objects that act in this way. So in a blackbody, some of the photons that go in never come out as incident photons. Such photons are destroyed or lost to heat. Most objects act as blackbodies in the sense that they give back a lower quantity of photons than the quantity received as a way of maintaining a certain level of heat. The blackbody effect serves as a kind of temperature regulator: it's the way nature helps objects avoid becoming too hot or too cold.

But with light-wave theory this cannot happen. A light-wave is a group of photons acting as one electric field. That is, a single energy field with multiple joined-up photons (a continuous wave), and no photon can act separately. Hence, in the context of the blackbody effect it's 'all or nothing'. Either the whole light-wave goes into an object and comes out still whole, or it does not come out at all. This contradicts the way the blackbody effect works. Light-wave theory cannot explain this contradiction.

5. Compton effect contradiction. When high-energy photons (i.e. highly concentrated photons) hit electrons it can release loosely bound electrons from their atoms. This results in a transfer of energy from these photons to such electrons. In so doing, the liberated and over-energised electrons shoot off a bunch of newly-created photons in quick succession, without any photon attenuation taking place. Such an effect is the so-called 'scattering' of photons.

Light-wave theory cannot explain the Compton effect because a light-wave is a single field of uniformly spread energy. Hence, a light-wave cannot have enough concentrated energy at its front edge so as to hit and release loosely bound electrons from their atoms. The Compton effect can readily be seen in the production of X-rays, in many gamma-ray spectroscopy experiments, in radiotherapy and in high-energy Compton cameras used in astronomy. Light-wave theory cannot explain such effects and hence contradicts the way the Compton effect works.

Contemporary physics is increasingly showing the spurious nature of light-wave theory. Powerful computer-aided spectroscopy clearly shows that the frequency of light is not derived from the electromagnetic oscillations of photons. That the calculation of the frequency of light is derived by breaking the light down to its colour spectrum and then arriving at a frequency measurement corresponding to its colour spectrum, and this in turn reveals the wavelength or distance between moving photons, and hence the concentration of photons in a given light ray.

When spectroscopy devices measure light frequencies, they're not measuring electromagnetic oscillations, but rather the colour spectrum. This, in turn, can reveal patterns of crests and valleys that are seen in spectroscopy devices, but these are entirely artificial patterns created by the spectroscopy device to help scientists better understand the results. The crest-to-crest distance represents the distance between moving photons.

The big misunderstanding of light is very widespread and entrenched in modern physics. The following quote is typical and is found in various guises in much educational material and scientific articles:

"The oscillations of the wave's electromagnetic radiation are characterised by rises and falls that form crests and troughs. The distance between the crests and troughs, together with the wave's speed of propagation determines the values of the frequency and wavelength, and the electromagnetic oscillations determine the energy of light."

Everything in the above quote is incorrect. The electromagnetic oscillations of photons are not characterised by a wavy line with crests and troughs.

Light moves as a sinusoidal vibration. The speed of light does not determine the frequency or wavelength whatsoever. And the electromagnetic oscillations of light have nothing to do with the energy of light.

In passing, it is mentioned that in 2015 the European Ecole Polytechnique Fédérale de Lausanne announced with great fanfare that light had been photographed as both particle and wave, thus proving the duality of light. Here is the photo:

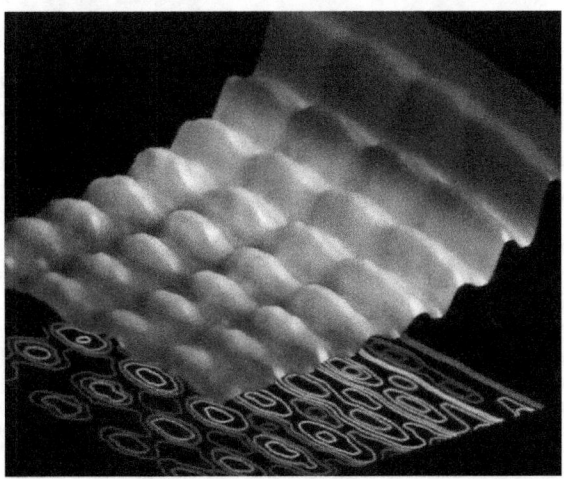

Later it was clarified that it was not a photograph of light itself. Rather, it was a photograph of 'surface plasmon polaritons' based on how electrons move on the surface of metals. There is no duality to light in the sense of light simultaneously having both wave and particle properties.

*

Do light-waves exist?

The short answer is no. Light-waves (or 'light waves' without a hyphen) do not exist at all, in any shape or form. This assumes a light-wave is defined as follows:

"Light-wave theory describes a so-called 'light-wave' as an electromagnetic wave or field containing multiple photons spread uniformly, with synchronised oscillations of electric and magnetic fields. The position of a light-wave within the electromagnetic spectrum is characterised by its frequency (speed) of oscillation. This frequency of its oscillation defines the light-wave's wavelength".

The above is how a light-wave is commonly defined in scientific literature. There are two possibilities: 1. Light either travels as one or more light-waves or 2. Light travels as one or more light rays (i.e. a stream of uncoupled photons). It cannot be that sometimes light moves as a light-wave and sometimes as a light ray. If you accept that light travels as a light-wave then you accept there is no big misunderstanding of light as explained in this book. If you accept that light-wave theory is spurious and that light-waves do not exist, then you are on the good path to understanding the magnificent and true fundamental nature of light.

Once you accept the spurious nature of light-wave

theory, many confusing and puzzling things about light fall into place and the big misunderstanding of light fades away. Whenever you read about or talk about the nature of light, if the word 'light-wave' comes into the text or conversation you should be very sceptical about the subject. 'Light ray' should be the common phrase instead of 'light-wave.'

If you try to get an answer to the question: *'how is a wavelength of light measured?'* you are told that it is the distance between two successive, equivalent points on a wave, for example from wave crest to wave crest.

Such a measurement has never been possible to carry out, or see, or prove in any way except as a theoretical concept. In wave theory, the distance (length) of the wavelength is inferred from the energy of light. The greater the energy, the shorter the distance between wave crests, and the quicker the oscillations of the photons.

The energy of light can be measured accurately in various ways, but mainly by its colour spectrum as seen in a spectroscope, and certainly not from any kind of theoretical wavy pattern.

The reality is that the greater the energy, the shorter the distance between the moving photons as they continue to oscillate at the same constant speed.

Here is a 6-word question you can ask anybody involved with the nature of light: *what determines the energy of light?* You and I know that the answer is: *the energy of light is determined by the concentration of photons in light.* If you get any other answer be sceptical about what you are reading or hearing.

*

What is the invisible spectrum of light?

The full electromagnetic spectrum contains various types of light. All these types of light (sometimes called 'types of radiation') contain just regular photons and nothing else. When reference is made to 'electromagnetic radiation', it is just another way of saying 'light'.

The only difference between the various types of electromagnetic radiation is the concentration (the number) of photons in such radiation. The concentration is determined by the distance between moving photons. If the distance is short, the concentration of photons and the energy of light in a given light ray will be high.

When the energy of a light ray is high enough to detach electrons from their atoms, such energy is usually referred to as 'ionising radiation'. Gamma rays and X-rays are examples of ionising radiation. Visible light, laser light, infrared, microwaves, and radio waves are examples of non-ionizing radiation.

Ionising radiation, like all types of electromagnetic radiation, is made of moving photons and nothing else at all. But the high concentration of photons in ionising radiation can overpower the electrons of atoms in an object and detach some of the electrons from their atoms.

But all photons contain a little energy. So when a high concentration of photons hit an object, the atoms in the object will absorb the energy from the barrage of incoming photons. This in turn over-energises the atoms, making them vibrate. This vibration causes heat in the object receiving ionised radiation from the photons.

So ionised radiation makes atoms move more quickly. As they do so, the atoms release their excess energy in the form of heat which can be harmful. For example, the heat from gamma rays or X-rays in excess can harm the body in various ways, such as radiation sickness, cancer, and organ damage.

Another example is when sunlight is focused on a piece of paper using a magnifying glass. This creates ionising radiation which produces enough heat to dislodge some electrons from their atoms, and also to burn the paper.

X-ray radiography can see into things (and human bodies) by using a variety of techniques such as differences in attenuation, or by photographing the ionising radiation as it leaves an object or body.

All types of heat are made of high concentrations of photons at a fundamental level. Thus when you hear that there are different types of heat arising from atoms, molecules and ions, this is so. But it all boils down to high concentrations of photons at a fundamental level.

If you are told that heat is the transfer of thermal energy, this is so. But the thermal energy that is being transferred is made of photons.

At the other end of the electromagnetic spectrum we have so-called 'radio waves' which are not waves at all. They are regular photons moving forward at the full speed of light but with a bigger distance between each moving photon. This distance can vary between a few millimetres to several kilometres.

Radio waves are typically generated by using intermittent electricity and a transmitter so as to produce photons at different intervals (different wavelengths). A little like morse code, each different wavelength conveys a different signal. These signals are received by a radio or receiver that translates the signals into sound, pictures and so on. Televisions, radios, smartphones, radar and many other devices all use radio waves to broadcast and receive signals through the air.

All radio waves are nothing more than moving photons like any other kind of electromagnetic radiation. The different wavelengths (distances between moving photons) are the key. Smartphones, radios, TV's, radio astronomy, and most kinds of telecommunications use radio waves. Such technology is designed to detect and use the different wavelengths in a variety of ways.

Microwave ovens work on similar principles. Microwaves contain a special internal light bulb called a magnetron that emits photons when the microwave is switched on. These photons are absorbed by the food/liquid in the microwave. In doing so, the photons are destroyed by transferring their heat to the food or liquid, and incident photons are emitted in their place. The cooked food or liquid emits fewer photons than the number of photons received and this maximises the heat retained.

Why can't we see the invisible spectrum? you might well ask. It's to do with our biological defences. Put another way, why is human vision limited to just a small section of the spectrum of light? Our photoreceptors are very sensitive to heat. The photoreceptors in the cones and rods behind the eyes are triggered by incoming light. But if the incoming light is too hot, such as ultraviolet or infrared light, the photoreceptors shut down to protect the delicate cones and rods. Equally, if the incoming light is too cold, such as radio waves, the lack of heat will not be enough to trigger the photoreceptors into action.

For example, let's suppose we are looking at a green leaf in the daylight. The green leaf will be continuously absorbing daylight photons and emitting incident photons in their place. The incident photons will be streams of moving photons with, say, a distance of 500 nm between any two moving photons. When these photons reach the photoreceptors, the 500 nm distance will determine the concentration (quantity) of photons coming in. This concentration of photons translates to a given amount of heat, and this amount of heat triggers the photoreceptors into providing the colour green. So each different concentration of photons gives a different temperature, which in turn gives a different colour. Each different temperature triggers a particular mix of cones that is instantly matched to a particular colour.

So the question is: *how can the spacing between moving photons translate to the colours we see?* And the answer is that the spacing determines the concentration of photons coming into the eyes. This concentration determines the degree of heat in a given group of photons, and this in turn translates to the colour we see.

By understanding the fundamental nature of light, it becomes much easier and quicker to understand or learn optical physics, electrical engineering, optical astronomy and many other branches of science associated with light. You will be streets ahead of your peers in such fields by not being blindsided and misguided by false light-wave theory.

*

How do we see colours?

In different parts of this book the way we perceive light and the way light moves has been discussed. But how do we actually see colours? Here are some typical (albeit erroneous) answers to this question.

Four typical statements giving erroneous information:

1. The 'colour' of an object is the wavelength of light that it reflects. This is determined by the arrangement of electrons in the atoms of that substance that will absorb and re-emit photons of particular energies according to complicated quantum laws.

2. Light is made up of different wavelengths, or colours, and non-incident light is a combination of all of them. For example, when white sunlight hits a blue patch on a beach ball, the blue patch reflects the blue wavelengths and absorbs all the others. The blue reflected light-waves from the Sun bounce off the beach ball, right into your eye. That's when the action starts.

3. When sunlight hits an object, some materials will absorb specific wavelengths. The wavelengths that aren't absorbed get reflected. This reflected light then reaches our eyes and makes us perceive the reflecting object as being a particular colour.

4. Any particular colour that we see in an object is determined by the wavelength of light that it reflects. For example, plants appear green because they

absorb all other wavelengths of light, leaving just green that is reflected (so it is green light that hits our eyes).

These four examples of how we see colours are wholly incorrect, yet such views are commonplace. You will know from reading this book that when light hits an object, the photons in the light are attenuated. That is, the photons are absorbed and gone forever, and newly created photons are emitted in their place.

These newly created photons are emitted by electrons but it takes an electron a *little bit of time* to absorb a photon and then create a new replacement photon (an incident photon). The mentioned 'little bit of time' inserts a tiny time-interval between each moving photon in a stream of photons.

This time interval determines the real physical distance between the moving photons, and this distance is referred to as the wavelength. So the length of a wavelength is determined by the real physical length of distance between the moving photons in a light ray going to your eyes.

Seeing colours

Now we come to how we actually see colours. The wavelength in a light ray determines which specific colour we see. Always remember that the wavelength is simply the physical distance between any two photons as they travel forward.

In reality, we do not receive just one line of photons

going into our eyes one after another, all equidistant, light marching soldiers. In fact our eyes are receiving many 'little' groups of photons simultaneously. Each little group is a stream of photons with a mix of wavelengths that gives the brain a specific recipe for matching the colour being looked at. So each little stream of photons gives the eyes and brain one or more wavelengths which instantly makes us see the colour we are looking at.

As you move around and see different things those little streams of photons will be changing the mix of wavelengths they carry, depending on what you are looking at, how you are moving and other factors. When seeing things, our cones are continuously being used and reused in different configurations.

The distance between the moving photons entering the eyes is what determines the colours we see. This distance/wavelength determines the energy of the incoming light rays and hence the heat signature of such rays. The eyes/brain detect that heat signature and this triggers the particular colours we see.

So at any one moment in time your eyes receive one or more light rays, each one giving you a colour recipe that's in accordance with the colours you are looking at.

The brain is very efficient at mapping out these many colours into the panorama of colours and shapes that we see. Furthermore, the brain remembers things, so by combining such memory with the incoming light rays, the brain provides a perception of 'instant vision'.

The crucial point to understand is that the physical distance between each of the moving photons is what determines the particular wavelength and hence the particular colour that we see at that moment in time. Nothing else determines which colour is seen. To emphasise this important point: only the distance between the moving photons determines the colour seen - nothing else affects the colours we see.

Seeing thousands of colours

Going on from this, there are many thousands of colour-hues that we see in our daily life. It is commonly thought that all the colours we see emanate from just three primary colours: red, green and blue. This is not quite accurate. The mentioned three primaries will give us any of the colour-hues we see, but cannot give us all the vivid (pure) colours we see such as pink and black.

It turns out that the colours cyan, magenta and yellow are just as much 'primary' colours as red, green and blue. *"We see things when light enters our eyes in two ways: (1) directly from a light source; and (2) emitted from an object. But the idea that the three primaries RGB can create all the colours in the world is totally false"* (source: Stephen Westland, Professor of Colour Science at the University of Leeds, England, Sept. 2023).

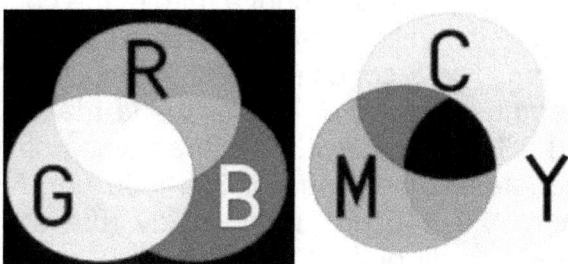

A discussion of how colours can be combined to make different colours and colour-hues, and how topics such as additive and subtractive colours work, can be very involved and is a little beyond the scope of this book.

The point here is twofold: 1. Red, green and blue are indeed the primary colours for all other colours including, cyan, magenta and yellow. 2. But to see certain vivid colours we also need cyan, magenta and yellow or some combination.

Another widely held misconception relates to our colour cones. We have nearly seven million colour cones behind our eyes which allow for the perception of colour. These cones are divided into three types that are designated long (L), medium (M) and short (S) cones. Each of the three types has a unique chemical makeup. The misconception is to think that each type of cone is specific to red, green or blue.

In fact, the L cones respond the most to light of the longer red wavelengths, peaking at about 560 nm.

The majority of the human cones are of the long type. The M cones make up about a third of the cones, and respond to the yellow-to-green wavelengths, peaking at 530 nm. The S cones respond the most to the blue wavelength, peaking at 420 nm, and make up only around 2% of the cones. The three types have peak wavelengths in the range of 564–580 nm, 534–545 nm, and 420–440 nm, respectively (nm stands for nanometre, a billionth of a metre).

Here is an image to illustrate the aforementioned graphically:

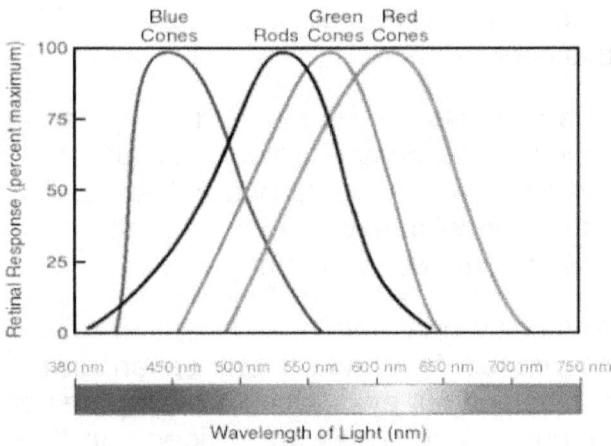

In this image we see that the sensitivity range of our three cone chemicals all overlap. So, for example, the brain may choose to use a red cone to help us see some hue of the colour green.

Here is another colour-related misconception that goes like this:

"There is no way to be sure that we all see the same colours. You learn from life that colours have names such as red or blue. But for all we know, you and I see an entirely different colour when we both look at the same colour".

This is a myth based on not understanding the fundamental nature of light. All humans have the same biology when it comes to perceiving colours through the eyes and brain. And by understanding how we perceive colours (as explained in these pages) it becomes clear that all humans perceive colours in the same way. But of course some people may have damaged eyesight or some inherited eyesight condition, in which case their perception of colours may deviate from the norm.

When we see the visible colour spectrum (e.g. the colours in a rainbow or prism) we colloquially say that we are looking at six colours because those are the colours that stand out the most: red, orange, yellow, green, blue, and violet. But in reality the visible colour spectrum contains thousands of colour-hues based on mixes of the six main colours. All these thousands of colour-hues are made from just these six main colours.

We perceive the colours white and black just like any other colour. These two colours are made up from other colours just like all the other composite colours we see. For example, stadium spotlights try to emulate white daylight. They do this by feeding equal amounts of red, green and blue light to a main spotlight to give very realistic white daylight for a football game. But if you were to mix red, green and blue paint in equal amounts, the result would be a murky brown-grey colour because of the unreliable wavelength accuracy of each type of paint.

And the colour black is made from cyan, magenta and yellow. If you hear that white and black represent the absence of colour this is not so. No colour or colour-hue (including white and black) can somehow carry other colours or carry other wavelengths of colour when moving forward as a light ray. A light ray is nothing more than a group of moving photons, and a photon cannot carry anything. Each photon is just a little packet of oscillating energy and nothing more.

You will know from reading this book that the distance between any two moving photons in a light ray is referred to as a wavelength. And the wavelength, normally measured in nanometres (nm), determines the colour. So every one of the thousands of colours and colour-hues that we see in our daily life has come from a light ray entering our eyes. And each light ray provides a mix of wavelengths (a mix of physical distances between photons) that gives the brain a 'light recipe'. This allegorical light recipe tells the brain which and how many light cones to throw into the cooking pot, i.e. what proportion or percentage of cones to use. Then in an instant the recipe is cooked and we are made to see the colour we are looking at.

The following image shows the range of physical

distances for each of the main colours. For example, if two moving photons are separated by a distance of 595 nm, then when those two photons enter the eyes they will convey the colour orange to the brain and you will see orange.

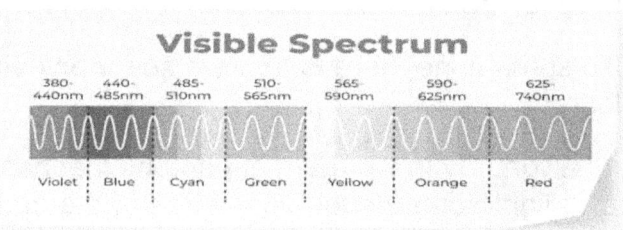

The nanometre distances representing different colours is not an exact science because a nanometre is such a small distance (a billionth of a metre). So, for example, when we see yellow it means we are seeing a distance between moving photons that could be anything from about 565 to 590 nm. To be clear when we see yellow, the arriving photons might have a distance of 573 nm or 584 nm or any other number that falls between about 565 and 590 nm.

Naturally, some of the incoming light rays will activate more cones than others, depending on the particular mix of wavelengths. Typically, we will have lots of different light rays entering our eyes simultaneously. And as each light ray is a 'little' group of photons with a mix of different wavelengths, each light ray will elicit the use of some of the cones. This is how we see lots of different colours simultaneously. But even so we would never be using all seven million cones at the same time.

"Humans have very sensitive perception of colour and can distinguish about 500 levels of brightness, 200 different hues, and 20 steps of saturation; in all, about 2 million distinct colours" (source: Vision - Transduction of Light, LibreText Biology, University of California, USA).

To summarise, here is how we see many different colours:

1. A light ray is a stream of photons that has come from a light-source such as a torch of the sun. Or it has come as a result of attenuation, i.e. as a result of light being absorbed and then emitted as new incident light.

2. Any light ray (however created or emitted) is entirely made of moving photons and nothing else at all. The only difference between various light rays is the physical distance between each moving photon in a given light ray. This physical distance is referred to as the so-called wavelength of light. So the length of a wavelength is the length of distance between two moving photons.

3. Any particular light ray will contain many moving photons, all emanating from a given source such as a light bulb, a torch, the Sun, etc. Or the light ray could have been emitted from any object such as a brown desk, a green plant, a blue beach ball, etc. Whatever the source of the light ray, every photon in the stream will be moving at the speed of light. But each light ray

will have its own particular distances (particular wavelengths) between its moving photons.

4. This means that there will be a given physical distance (a given wavelength) between any two photons in a light ray. When this light ray enters the eyes, the various wavelengths in the light ray will trigger some mix of cones so as to instantly give us the colour that at that moment we are seeing. This mix of cones determines the proportional mix. Seeing any colour is not about the quantity of eye-cones used in the light recipe, it's all about the proportion (percentage) of eye-cones used in the light recipe.

5. Nearly all the colours that we see in our daily lives will be colour-hues rather than one of the primary colours. For example, even if we think we are seeing pure green when looking at a green leaf, we are in fact seeing a colour-hue of green, not pure green. So just about every light ray will be a light-recipe for a particular colour-hue. Whatever the colour-hue, the eyes and brain will instantly cook up a colour recipe for the given colour-hue by just using some combination of the three primary colours.

6. Here are three examples of a light ray. The string of numbers in each example represents a string of distances (the wavelengths) between each moving photon in a light ray.

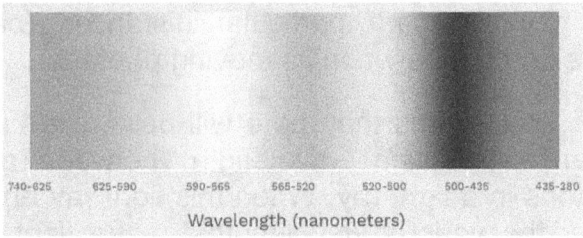
Wavelength (nanometers)

1. Example of a magenta light ray: 11111311211113311112111. In this example it is assumed the hue of magenta is made from a proportional mix of violet, yellow and green. So number 1 is a distance of, say, 401 nm, 3 is a distance of 590 nm, and 2 is a distance of 500 nm. Each number in this string of numbers is a wavelength (a distance between two arriving photons). So in this example, the brain will instantly create a recipe with a predominance of 401 nm wavelengths and a small amount of 590 nm and 500 nm wavelengths to give the precise hue of magenta being looked at.

2. Example of a pure white (incident) light ray: 664654455546645. In this example, number 6 is 650 nm (red), 4 is 450 nm (blue) and 5 is 550 nm (green). The brain will instantly create a recipe for making you see pure white because the proportions of red-green-blue are equal in this example. But suppose there were two extra 6's in that line of numbers. You would have seven 6's, five 4's and five 5's. This would still give you white, but with a slight red tinge.

3. Example of a light ray from seeing a wooden floor: 44546566665564666. In this example we are looking at a wooden floor which is brown. The colour brown is made from red, green and blue, but with more red than green or blue. So in this line of numbers, 4 gives 450 nm (blue), 5 gives 550 nm (green) and 6 gives 650 nm (red). In looking at the wooden floor you may also see a few streaks of very dark brown. Thus you are simultaneously looking at two different shades of brown. When that happens it means that at the moment of seeing both shades of brown your eyes are receiving two separate light rays: one for light brown (as in this example) and one for very dark brown with a different set of numbers for a dark brown light-recipe.

To summarise, at any given moment, our eyes can receive many little groups of photons. Each little group is a separate light ray of moving photons giving the eyes a particular series (mix) of wavelengths. This mix of wavelengths is 'put together' into a particular light recipe by the brain. This final light recipe is the final actual colour we see. The key point here is that the wavelength determines the colour, and this wavelength is nothing more than the physical distance between any two moving photons.

By human convention, the following six colours represent the visible spectrum:

> **Violet: 380–450 nm**
> **Blue: 450–495 nm**
> **Green: 495–570 nm**
> **Yellow: 570–590 nm**
> **Orange: 590–620 nm**
> **Red: 620–750 nm**

Apart from these six colours, no other colours have a designated wavelength (a designated distance). This criterion is based on the colours that we humans are able to see in the visible spectrum or in a rainbow. Thus, as we can only see these six colours, we give each colour a specific wavelength which reflects or relates to the distance between two moving photons. No other colour or colour-hue is given its own wavelength because they are composite colours, they are colours made from a combination of the primary colours red-green-blue.

The colours violet, yellow and orange are given their own designated wavelengths in spite of being composite colours, simply because these three colours stand out in the visible spectrum.

But if you are looking at pink for example, the light ray going into your eyes will not have a human-designated wavelength. Pink is made from red and violet, each colour at opposite ends of the visible spectrum. So the distances between the photons in a light ray travelling from say a pink wall to your eyes will look something like this: 400- 400- 400- 400-700- 700- 700- 470. The repeated 400 and 700 nm distances give a predominance of violet and red, plus a little blue (470 nm) because this is required for the particular pink-hue of the wall we are looking at.

Thus when daylight is attenuated by the pink wall, the electrons in the pink wall will emit photons with a variety of wavelengths that when received by the eyes will instantly make you see pink. This is how the human perception of colours works for any colour or colour-hue that is not one of the six colours with designated wavelengths.

To finish this section, a question that arises goes like this: We have said that when photons go into the eyes this triggers various combinations of eye-cones to make us see certain colours. But how do the incoming photons trigger the right combinations of eye-cones for the colours we are seeing?

The answer is heat. As mentioned, the eyes do not perceive the photons themselves, as if coming in like marching soldiers with different spacings between the soldiers. But the eyes, cones, rods, etc. do perceive heat to which they are very sensitive. And as explained previously in the book, the distance between moving photons determines their level of energy, and hence their level of heat. Every little group of photons going into the eyes will be giving off a particular heat signature which in turn triggers the colour to be seen.

*

How do we see a specific colour?

In the previous section we discussed how we see colours generally. We also saw how an object or material reveals its colour. How it emits a ray of light with a mix of wavelengths that tell our eyes what colour we are seeing.

But the following question arises: *How can a given object or material know what particular mix of wavelengths to emit so as to convey a given colour?* In other words, what makes an object, such as a blue beach ball or a pink wall send out a light ray containing the exact combination of wavelengths that will make us see the colour of the object that we are looking at?

This question has intrigued optical scientists for many years. A common answer to this question is usually based on the big misunderstanding of light as explained in this book. Such an answer goes something like this:

"When for example daylight (which is a combination of all wavelengths) hits an object, some objects will absorb specific wavelengths. The wavelengths (i.e. the light-waves) that aren't absorbed get reflected. This reflected light (i.e. the reflected light-waves) then reaches our eyes and makes us perceive the reflecting object as being the particular colour that we see".

Quite apart from the fact that light can never bounce or reflect off anything, the problem with this answer is that it does not answer the question. It does not tell us exactly how an object conveys its colour to human eyes. The implication is that the non-absorbed light-waves that bounce off an object are somehow mysteriously imbued with the exact colour mix of said object (the exact proportions of each colour in the mix of the colour-hue). And that having been imbued with the correct colour mix, such non-absorbed light-waves are then able to convey this colour mix to the brain.

Some physicists entrenched in light-wave theory claim that light-waves are indeed fully absorbed into electrons (not reflected), and that such electrons will then emit new whole light-waves. But ask the question: *how exactly do electrons emit new whole light-waves?* and there is no comprehensible, credible answer. Some may say vaguely that the answer lies in a set of very complicated equations. It's probably best if we do not continue down that rabbit hole.

Back to reality. To explain exactly how an object conveys its colour to our eyes we will use the example of a pink wall. So if looking at a pink wall, what makes the pink wall emit a particular mix of wavelengths for red, violet, and a little blue so as to convey the exact colour-hue of pink that we are looking at?

We need to understand how the pink wall is able to send out a light ray containing a precise mix of wavelengths (distances between moving photons) that exactly conveys the pink hue of the wall. To understand how this works we first need to discuss valence electrons.

The valence electron

All atoms have electrons spinning around the nucleus. Some electrons spin around in higher orbitals than other electrons in the same atom. The electrons farthest away from the nucleus of their atom are called valence electrons.

All atoms have 7 possible orbital levels. By default valence electrons stay at orbital level 2 until/unless energised by an incoming photon, in which case they jump up to a higher orbital level. Non-valence electrons stay below level 2 and rarely if ever emit photons. The following image (courtesy of the Khan Academy, USA) shows this graphically:

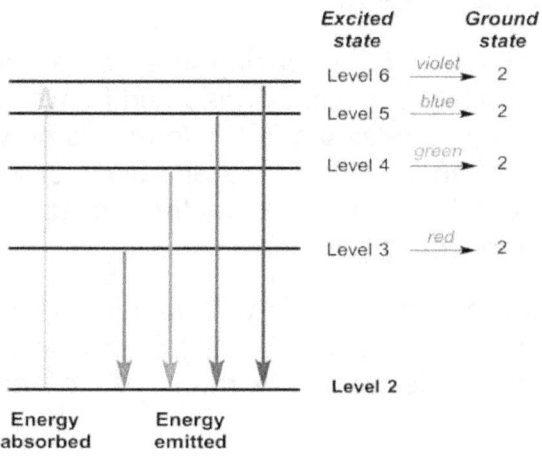

This image shows for example that if an incoming photon hits a valence electron in the atom of a red-coloured object, the valence electron will absorb the photon's energy making it jump up from level 2 to level 3. It will then release a new photon and fall back to level 2. If the object is a blue colour, the electron will jump up to level 5 and then fall back to level 2, and so on for other colours and levels.

All valence electrons stay at their default level 2 (their ground state) unless energised. So whenever a photon hits a valence electron, the electron will absorb the photon's energy and be super-energised.

By being super-energised the valence electron will then jump to a higher orbital level around its atom. But in doing so, the electron becomes a little unstable and is at risk of breaking away from its atom completely. Allegorically, the super-energised electron becomes over-excited and nervous and wants to go back down to its default level 2.

It does this by releasing the same amount of energy that was originally absorbed from the photon. This energy is released in the form of a new photon, and the valence electron is then able to drop down to level 2 and be ready to repeat the process if it meets another incoming photon.

Note: below level 2 you have so-called 'core electrons' that act to protect the inner core of the atom. Core electrons are rarely excited enough to emit photons unless extreme temperatures overwhelm the atom's valence electrons.

So when a valence electron releases a new photon, for example from a red-coloured object, the incident photon will go shooting out of the atom. In reality millions of such photons will be shooting out of millions of atoms in the red object. So wherever you are standing nearby you will be able to see the red object by virtue of receiving the incident photons at any angle.

Some atoms have more valence electrons than others (anything from one to eight valence electrons). For example, atoms in red objects will normally have one valence electron. At the other end of the colour spectrum, the atoms in violet objects will normally have 7 valence electrons. So different coloured objects can have different amounts of valence electrons in their atoms. A periodic table of the elements lists the number of valence electrons for every known element. And by knowing the colour of the element you can find out the number of valence electrons per atom for just about any object.

Valence electrons help maintain the stability of an atom so that it does not become overloaded with energy. They do this by pumping out incident photons, one after another in quick succession for as long as photons keep coming in. The rate at which photons can be pumped out of an atom depends on the number of valence electrons in said atom. In science this is called 'spontaneous emission'. Most atoms have more than one valence electron, so several valence electrons could be emitting photons within the same atom. This is a well-studied and well understood subject.

The greater the number of valence electrons in an atom the quicker the emission of photons. That is, an atom with more valence electrons can pump out photons at a quicker rate than an atom with fewer valence electrons.

This is so because in an atom with multiple valence electrons, each electron can absorb and emit photons independently of other electrons in the same atom. Furthermore, valence electrons within the same atom can emit photons simultaneously, or sequentially one after another.

Coming back to how an object (such as a pink wall) conveys its colour to our eyes, here is a point by point explanation.

1. We are looking at a pink wall in the daylight (pink is made from the colours red, violet and blue). The daylight in the room is absorbed into the wall. So incident photons are emitted out of the wall and when they reach our eyes we see the pink colour of the wall. These incident photons travel out from the wall as many 'little' light rays going in all directions. So we can see the same colour pink of the wall from just about any angle.

2. Every one of these little light rays will carry the same mix of wavelengths so as to convey the same colour pink to anybody else who may look at the wall. This mix of wavelengths is specific to the particular pink wall and stays fixed (unchanging) indefinitely unless the wall is changed in some way. Thus any object, a white cup, a black hat, a red sofa, will be sending out a fixed, unchanging mix of wavelengths until the object is changed, moved, etc.

3. Even though the pink paint is dry, it has a mix of red, violet and blue molecules tightly packed together under the surface of the pink paint on the wall. This mix of molecules is what makes the colour pink. These molecules of different colours stay intact as separate colours, even though you would not think so when looking at the single colour pink on the wall. Each molecule will have two or more atoms, and each atom will be attenuating photons and emitting incident photons.

So for example, the atoms in the red molecules under the pink paint will be emitting one photon at a time from its solitary valence electron. This valence electron then has to 'recover' by going back down to level 2. It can then start over again when the next photon comes calling. This puts a relatively long time interval between each emitted photon, resulting in a distance of say 720 nm between each moving photon emitted by the mentioned valence electron.

For violet molecules, the atoms will be emitting many photons in quick succession by virtue of having

7 valence electrons. This results in a distance of say 412 nm between the emitted photons. And for the blue molecules in the pink wall, the atoms would typically have two valence electrons, allowing an emission of photons with a moving distance of say 493 nm (slower than violet, but quicker than red).

4. So the pink wall will be emitting many millions of photon-groups in all directions, with a mix of wavelengths corresponding to 412 nm (violet), 493 nm (blue) and 720 nm (red). First out of the wall will be the photons for 412-violet, followed by 493-blue and then 720-red. They go out in this order simply because of speed. The smaller the physical distance between each moving photon, the faster the journey-time of the whole light ray. Note that all photons move at the same speed of light, but of course the journey-time of a whole specific group of moving photons will vary.

5. When the photons are emitted they will go out of the pink wall in a repeating sequence. And this continues indefinitely for as long as there is light in the room with the pink wall. Furthermore, the number of photons for each colour will be proportional. For example, the light recipe (the mix of wavelengths conveying the colour pink) might be: 412-412-412-412-493-493-720. This tells the eyes not only which light cones to trigger, but also how many cones of each colour to trigger so as to give the exact colour-hue of the pink wall.

As mentioned, the streams of photons giving 412-412-412-412-493-493-720 will be a repeating sequence (a repeating colour recipe, a repeating mix of wavelengths). So there will be millions of streams of photons radiating out of the pink wall in all directions. Each stream will contain the same repeating sequence of a pink colour recipe that goes into the eyes.

The streams go out of the wall in all directions because when valence electrons emit photons, they do so while on-the-move as they whizz around the atom, and once a photon is emitted it can only travel in a straight line. But the repeating colour recipe that is emitted is 'set in stone' forever (never changing). The only thing that could change the repeating colour recipe would be some kind of change to the pink wall.

Thus, on any day that you walk into the room with the pink wall, you will see the exact same colour pink from any angle in the room, on any day. And your eyes will be continuously receiving the repeating colour recipe until you look at something else.

A final question to clarify is how does a light ray entering the eyes trigger the specific colour that we see? As mentioned, we have three types of eye cones: long wave, medium wave and short wave. Any particular colour or colour-hue that we see will be a mix of one to three of these cones. The mix, i.e. the proportion of cones from each of the three types, is determined by the mix of wavelengths contained in the incoming light ray.

If an incoming wavelength (into the eyes) is long, it means there is a long distance between moving photons, and this triggers the long wave cones into action. The same for medium and short wavelengths coming into the eyes. For example, if the wavelength is 500 nm, this distance of 500 nm between the incoming photons will trigger some medium wave cones in the eyes and tell our brain to make us see green.

It's not that the eyes and brain measure the distances between incoming photons. But those distances determine the degree of heat (energy) of the group of incoming photons. And that in turn determines the colour or hue that is seen. Each incoming ray of light has its particular 'heat signature' which determines what is seen.

Biology textbooks describe the aforementioned as follows:

"Exposure of the retina to incoming light hyperpolarizes the rods and cones, removing the inhibition of their bipolar cells. The now-active bipolar cells in turn stimulate the ganglion cells, which then send action potentials along their axons (which leave the eye's optic nerve). Thus, the visual system relies on changes in retinal activity to encode visual signals for the brain".

So now we know how an object tells the brain what colour to see. In short, the number of valence electrons in the atoms of an object determine the distance between each moving photon that is emitted. And these physical distances determine the wavelengths, the heat signature, and hence the colour of the object that we see.

No actual colour as such exists in nature, not even red, green or blue. But distances between moving photons do indeed exist in nature, and those distances are not coloured in any way at all. But those distances (the wavelengths) tell the eyes and brain what colour to see in our mind. Photographic cameras are designed to be similar; when streams of photons go into a camera, the wavelengths 'tell' the camera what colours to see and record as a photograph or video. In other words, the camera senses the heat of the incoming photons, referred to as 'thermal imaging'.

Note: This book has described how the nature of light makes us see colours, but for the detailed biology of vision the reader is urged to investigate the phrases 'visual system' and 'colour vision' in sources such as the internet and libraries.

*

How does a prism work?

The attenuation of light has been mentioned at various points in the book. Attenuation refers to the absorption of photons into the electrons of an object or medium followed by the emission of newly created photons that in effect replace the absorbed photons. It takes a moment of time for an electron to absorb a photon, emit a new photon and then be ready to repeat the process. As a result, the newly emitted photons will travel out of an object or medium with a given time interval between each moving photon.

The extent of this time interval is determined by the atomic composition of the object, medium or material. So different objects or mediums will put different time intervals between the emitted photons. This time interval determines the distance between each moving photon. And this distance determines the so-called 'wavelength' between every moving photon in a light ray.

The emitted photons from an object or medium are called 'incident' photons because the act of attenuation puts the mentioned time-interval (i.e. a physical and real distance) between each emitted photon in a permanent fixed manner. In other words, the distances put between each emitted photon are fixed forever until and unless such photons are destroyed or the object emitting the photons is changed.

To summarise this point, when light is attenuated into an object or medium, the light coming out of the object or medium will be new immutable incident light rays. These incident rays will have a fixed permanent distance between the moving photons in the light ray.

Everything that we see in our daily lives is made possible to see as a result of receiving incident light rays from any object or medium that we look at. Hence, at any moment in time, our bodies will be emitting millions of incident light rays, and also millions of incident light rays will be moving towards you. Of course, our eyes only receive a tiny fraction of those millions of light rays as it depends on what we are looking at.

All incident light rays are said to be coherent because they contain a specific mix of unchanging wavelengths (i.e. a mix of unchanging distances between photons). This has to be so otherwise different people would see a different image/colour of the same object or medium. Or the same person would see a different image/colour of the same image from day to day.

Now we come to incoherent light. When light is created in a light bulb, a torch, the sun, a lamp, flames, a star, candlelight, infrared emitter, ultraviolet, X-rays, etc. the light or radiation from such sources is said to be incoherent. The light is incoherent because it has a disorganised mix of all wavelengths. In other words the light has a disorganised mix of many different physical distances between its moving photons.

Incoherent light is created from extreme heat. Such heat over-excites atoms and makes the electrons in the atoms 'spit out' photons every which way in a mix of disorganised wavelengths. As such, incoherent light is non-incident light.

Technically, incoherent light is said to be unpolarised light because it is disorganised. And incident light is said to be polarised light because it is organised light with a given fixed mix of wavelengths. Any kind of incoherent light, whatever the source, is nothing more than moving photons, albeit moving with different, disorganised distances between the photons. The photons in coherent and incoherent light are identical and all move at the same constant speed of light 'c'.

Any incoherent light that falls within the visible spectrum is colloquially referred to as 'white light' which can cause confusion because some kinds of white light can be coherent. There is a common misconception that such white light somehow carries all the colours of the visible spectrum. This misconception has arisen mainly because if you shine white light such as daylight through a prism it appears to split into all the colours of the rainbow as in this image:

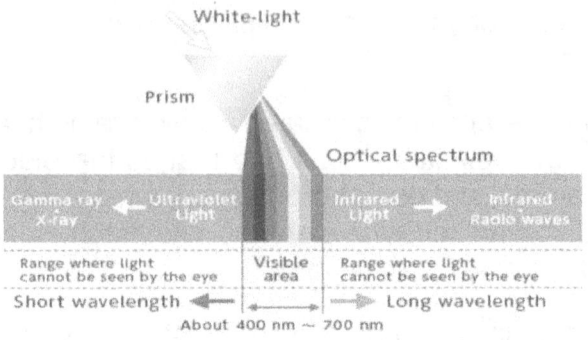

    Here we see that white light going through a prism apparently splits into the colours of the visible spectrum. It is often said that white light carries all the colours of the rainbow or that it can be split into the colours of the visible spectrum, but this is not so. White light does not split into any different colours - in fact white light does not split into anything at all. And white light, like any other kind of light or electromagnetic radiation, is just moving photons - nothing else at all.

    So what is happening? When white light hits the face of a prism it means that a big mix of different (incoherent) wavelengths hit the prism and are attenuated. The white light that is attenuated is destroyed and new incident light is emitted.

    The incoherent nature of white light means that a big mix of all the wavelengths of the visible spectrum will be hitting the face of a prism (or hitting our eyes if looking directly at the white light).

But even though the white light is incoherent, we know that it must contain a more or less equally divided mix of red, green and blue wavelengths. That is, it must contain a mix of distances falling somewhere into each of the three nanometre groups: 620–750 (red), 495–570 (green), and 450–495 (blue). If this were not so we wouldn't see the white light. The mix of wavelengths in white light does not have to be exactly divided equally into the three primary colours, just approximately.

Most of the photons in sunlight (but not all) are refracted through the Earth's atmosphere even on a cloudless day. There are many subtle hues of white light depending on weather conditions, amount of clouds, time of day and so on. But regardless, when we see white light it mostly just looks white.

But even though sunlight is mostly refracted, it remains mostly incoherent because the atmosphere (a medium) is very unstable and changeable from minute to minute. It is estimated that sunlight or daylight arriving on Earth has wavelengths (distances between moving photons) that vary between 200 nm and 2,500 nm, which more than fully covers the whole visible spectrum.

To clarify the terminology, coherent light and polarised light mean the same thing. They both refer to incident light rays with a fixed immutable mix of wavelengths that give the eyes or a spectroscope a given colour.

Equally, incoherent light and unpolarised light mean the same thing. They both refer to non-incident light rays with no fixed mix of wavelengths, just a disorganised random mix that gives the eyes or a spectroscope a white colour by virtue of having all the mixed wavelengths of the visible spectrum.

If you try to check the difference between coherent and polarised light (or incoherent and unpolarised light) using the internet and other sources you will quickly come up against light-wave theory. This will take you down confusing rabbit holes with talk of waves being in or out of phase, being filtered, oscillating in certain directions, different planes of vibration, stationary interference, and so on. Given the spurious nature of light-wave theory and the big misunderstanding of light, these aforementioned terms are meaningless. Don't be fooled or misled.

Coming back to prisms, there are many types:

Prisms find use in several fields like ophthalmology, optical instruments, and architecture. They are commonly seen in telescopes, binoculars, submarine periscopes, microscopes and some industrial uses such as spectroscopy and laser lights.

Commercially available prisms normally have some type of dispersive coating designed for the particular use of the prism. The following image (from left to right) shows a prism designed for showing the visible spectrum to maximum effect:

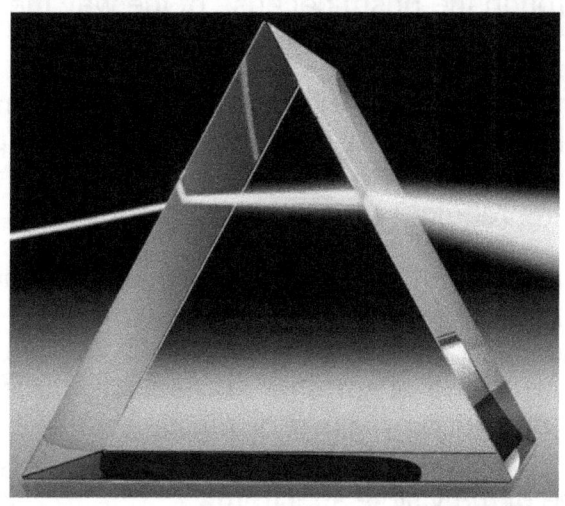

"In optics, a dispersive prism, which usually takes the shape of a triangular prism, is an optical prism that is used to disperse light into its spectral components, such as the colours of the rainbow" (source: Dispersive prism, Wikipedia.org).

To accomplish such dispersion, certain elements such as N-BK7 glass, flint, silica, and others are ground into a fine powder and stuck to the faces of the prism to provide a dispersive coating. The coating recipe can vary according to the type of prism being designed.

The dispersive coating is then carefully polished and calibrated to give the required translucency. So when white light hits the face of a prism the photons are absorbed into the dispersive coating and then new photons are emitted into the prism. They go into or through the prism because of the way the prism is designed, and also because of Snell's Law.

Prisms are made of high quality transparent plastic or glass and every step of producing a prism is carefully calibrated to a very high precision so as to give the required effect of light dispersion.

If the prism had no kind of dispersive coating and only offered very good quality transparency, the light would just go straight through with virtually no change. If the prism offered mediocre transparency from poor quality glass or plastic (but also had no dispersive coating) it would just give a very diffuse image of the visible spectrum.

So when white light is made to shine through a prism with a dispersive coating, the big mix of many different wavelengths in the white light will be attenuated through the face of the prism. All the white light-wavelengths falling into the red spectrum (from 620 to 750 nm) will attenuate into the prism as a separate ray of red light. The same for orange (from 590 to 620 nm) and so on, so that you end up seeing the whole visible spectrum going through the prism for as long as white light continues to shine at the prism.

Here is an image (from left to right) showing the way each colour stacks up starting with violet (always in the same sequence) as the colours go through and out of a prism:

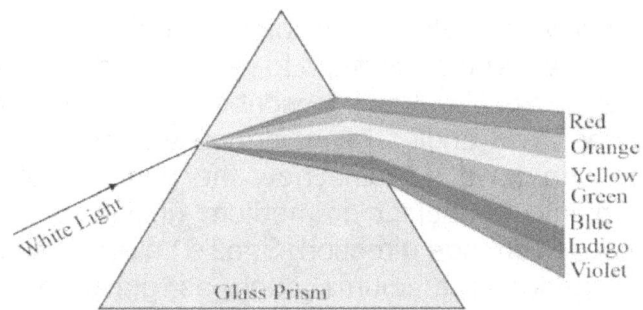

Each colour going through a prism is simply a stream of photons and nothing more. For example when we look at the colour violet going through a prism we are looking at a moving stream of photons with a mix of wavelengths for violet. We can see the violet ray because that same actual ray is sending a stream of photons to our eyes with a mix of distances between the moving photons. But all such distances fall within the 380-450 nm range for violet.

Also, when the white light was refracted, the quickest photons to be attenuated into the prism were those white light photons with the shortest wavelengths, i.e. the violet wavelengths. So all the white light-wavelengths falling between 380-450 nm were attenuated the quickest and emitted inside the prism as a light ray with a mix of wavelengths within the 380-450 nm range for violet. The second quickest was indigo and so on. This explains how the colours always stack up in the same sequence. The incoming light into the prism forms into distinct rays of colour that correspond to the speed of attenuation (refraction) for each colour.

As to what makes each colour go through the prism at a certain angle, we need to refer to Snell's Law. Put simply, Snell's Law describes the relationship between the angles of incidence and refraction. In other words, it explains how the angle of photons going into an object or medium (including prisms) appear to change direction. Snell's Law works best in homogeneous mediums such as glass, a prism, water, gas, etc. The angle of incidence and refraction is caused by gravity.

Given that photons can never change direction or bend, what happens is that the photons are attenuated (destroyed), and the emitted photons immediately go in their newly emitted direction inside the medium or prism, giving the illusion that the photons have changed direction. This new direction sends the photons towards an imaginary perpendicular line to the horizon of an object. This imaginary perpendicular line is called a 'normal'. For more information on this subject see 'normal force' and 'Snell's Law' in Wikipedia.org.

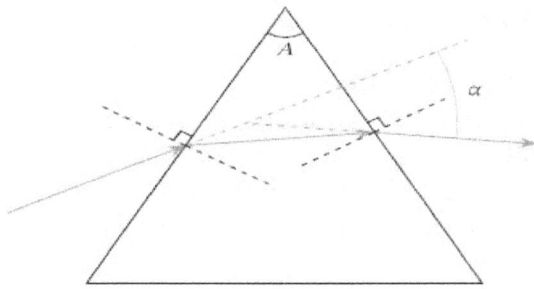

In this image the triangle represents a prism and the solid arrowed line going from left to right represents the path followed by light as it enters and leaves the prism. As the light enters the prism it is refracted. And as the light leaves the prism it is refracted again.

Snell's Law says that as the light goes into an object or medium such as water or a prism, the light will angle downwards a little towards the normal. And as the light leaves, it will take a new angle in relation to the second normal. The 'normal' are the two dotted perpendicular lines at right angles to each face of the triangle in the above image. Snell's Law is well understood and accepted in physics.

We have mentioned that light-wave theory is deeply rooted in contemporary physics. This was explained in the section *'What is the big misunderstanding of light?'* As a result, almost any type of research or study into the nature of light is bound to run into misconceptions and erroneous information about light. Here are three examples:

Example one

"As it passes through the prism, the white light is divided into its component rainbow colours. This division of white light into its different colours is known as dispersion. The prism slows down light, bending its path through the process of refraction. Each colour is caused by a different wave frequency (different rate of electromagnetic oscillations). "*These different wave frequencies cause the colours of light to bend at various angles as it passes through a prism."*

Everything in the above quote is incorrect. White light does not split into the colours of the rainbow. The colours of a prism are not determined by a 'wave frequency'. Photons of light never slow down. The frequency of light has nothing to do with electromagnetic oscillations, and light can never bend under any circumstances.

Example two

"The recombination of light is the phenomenon by which a ray of white light, once split, goes through a process that transforms it back into a ray of white light. That is, one process is the opposite of the other."

The above statement is false, and the following image is also false:

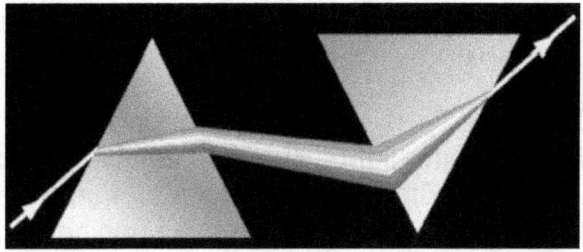

Any light that comes out of a prism, whether the prism is inverted or not, always comes out in the form of attenuated light. That is, such light is absorbed and new incident light is emitted, without any form of so-called 'recombining'. If the leaving face of a prism offers high-quality transparency, the light will exit largely unattenuated (as is). So there are no circumstances in which the colours inside the prism can be recombined into white light upon exiting.

The following image shows what really happens, all according to Snell's Law:

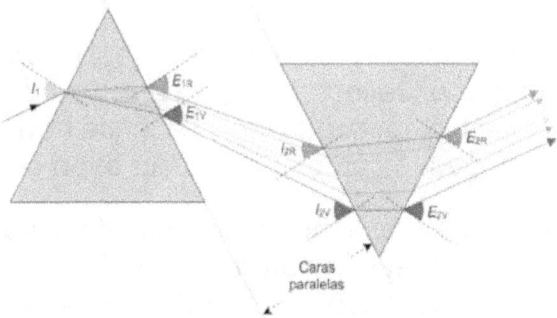

In this image (showing in Spanish two parallel prism faces) we see correctly that the light leaving the inverted prism is not recombined white light. It is light that was attenuated through the leaving face of the inverted prism.

Of course, it is possible to force light to leave a prism as white light. But this involves a complex arrangement of concave glasses, mirrors, special prisms and calibrated lighting & spacing. And even then, the outgoing white light would not be any kind of light recombination, it would just be newly attenuated (newly created) light that looks white.

Example three

Here is another false example: *"When white light hits the surface of a coloured object, a part of the visible spectrum is absorbed by said object. The unabsorbed light is reflected into our eyes to give us the colour we are looking at. Those reflected light-waves are what determines the object's colour. The colours we see are therefore those that the object reflects instead of absorbing"*

This example is also entirely false. No type of light can bounce or reflect from an object. And even if it were true, it is not explained in science how reflected light manages to capture the colour of an object.

The truth is that light-wave theory does not include the concept of absorption and emission of light from objects, so by necessity it offers false alternative concepts. Wikipedia correctly states the following:

"Light-wave theory... is incapable of explaining the phenomena of absorption and emission of light by

matter, or the interaction between matter and radiation" (source: Wikipedia.org).

It is lamentable that such misconceptions about the nature of light are so widespread in science and continue to be so commonly taught.

To finish on the subject of prisms, here is a unique photograph of a strand of spider-web silk:

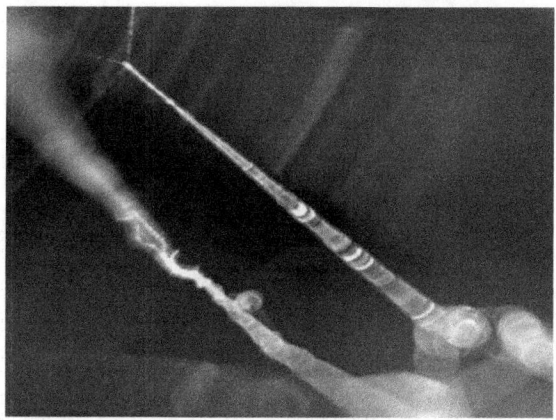

Photograph courtesy of Oswaldo (Owi) Ponce

Spider silk is generally round, semi translucent and semi hollow. It is made from an aberrant mix of amino acids. So when sunlight hits a silk thread, the light refracts into the thread. Then the refracted photons of light hit different amino acids along the length of the thread. As this happens, the light again refracts into and out of each amino acid. So we are looking at different colours of light emitted out of each different amino acid and then leaving the spider silk as a mix of light rays towards our eyes. A wonderful prism of nature.

*

Is the quantum theory of light correct?

The quantum theory of physics (same as the quantum theory of mechanics) was originally proposed by Niels Bohr and Max Planck. Later Albert Einstein proposed quantum theory in relation to light propagation specifically. Inspired by 'Planck's Law', Einstein proposed that light travels as discrete bundles of energy (each bundle later became known as a photon). Up to this point Einstein was correct, but Einstein then went on to say that each photon carries a quantity of energy equal to the product of the frequency of vibration of that photon, calculated with Planck's constant.

By 'frequency of vibration' Einstein is referring to the rate of electromagnetic oscillations of a photon - this is incorrect because it is saying that the rate of EM oscillations determines the energy level of a given photon, thus bestowing the photons themselves with different levels of energy. This incorrect proposal by Einstein went on to cement a belief in light as a light-wave, and a belief in the particle/wave duality of light. In short, the quantum theory of light led to the big misunderstanding of light as explained in this book.

Unfortunately, the quantum theory of light of the 20th century has morphed into the quantum physics of the 21st century. Light-wave theory is considered by many to be the basis of contemporary physics when it comes to subatomic particles. That is, quantum physics is firmly wedded to the belief that the electromagnetic oscillations of photons determine the energy and hence the frequency of light. This is totally spurious - there is no milder way to say it.

There is no attempt to disparage quantum physics or its adherents. No doubt eminent scientists such as Bohr, Planck, Einstein, Broglie, Huygens and others have contributed in good faith to the development of quantum physics. Furthermore, quantum physics has led to things like lasers, light-emitting diodes, transistors, medical imaging, electron microscopes, and a host of other modern devices.

But none of these mentioned technical developments and inventions use or depend on light-wave theory. Whether inspired or not by light-wave theory, these technical developments are said to come from 'quantum physics' simply because it is the generic umbrella name we give to the physics of subatomic particles.

So although quantum physics is fundamentally flawed regarding the nature of light, it has nevertheless served us well. It is hoped that with time this flawed concept of light will be recognised, thus putting quantum physics onto firmer ground and galvanising forward our knowledge of subatomic particles.

Albert Einstein, considered to be a pioneer in quantum theory, always remained sceptical on the subject. He famously said *"God does not play dice"* in reference to the inherent uncertainty in particle physics. Indeed, many contemporary physicists such as Roger Penrose and others think that quantum mechanics is simply baseless.

We humans are not capable of applying deterministic laws of physics to subatomic particles because the particles are so very tiny and they move so very fast. We cannot determine with any precision where/how a particular particle is going to move, so we rightly apply probabilistic laws (guess work) to subatomic particles - it's the best we humans can do given our current level of science.

But given that much human endeavour in particle physics is based on probability and uncertainty, it does not follow that we have to accept the irrational and contradictory nature of quantum physics such as the following:

1. That light is sometimes a wave with multiple connected photons, and at other times it can be streams of separate photons (the spurious 'duality of light').

2. That if particles are not being observed, then such particles cannot exist as either a wave or a particle. If this is so, then what form of existence are such particles enjoying while not being observed?

3. That double-slit experiments show how some photons go through both slits simultaneously, i.e. that a single photon goes through both slits at the same time. Then after this, the two split photons join up again into a single photon to help form a single wave

on the other side of the two-slit partition. This is said to help explain how so-called light-wave interference occurs so as to form a single wave that heads towards the photon detector screen.

4. That the act of observing subatomic particles changes their behaviour for as long as they are observed. Clearly, the physical act of arranging apparatus to observe particles can disturb and change their behaviour, but it is absurd to claim that such changed behaviour continues for as long as they are observed. In other words, that the act itself of being observed is what changes their behaviour rather than any physical disturbance of laboratory equipment.

5. That a group of particles, such as two photons or two electrons, can affect each other even if separated by a large distance. So-called 'quantum entanglement' refers to a phenomenon in which, say, two photons continue to interact and affect each other even if each photon is separated from the other thousands of miles apart. That somehow the two photons communicate with each other instantaneously, regardless of their distance from each other.

6. That subatomic particles can be in multiple places at the same time, i.e. that a given particle can be in several different places at once. This is called 'quantum superposition'. But the theory says that if you try to observe this phenomenon, the

superimposition disappears such that you can never actually observe it.

7. That light is both a wave and a particle phenomenon at the same time. Pilot Wave Theory (also known as the Broglie–Bohm theory) postulates that the particle/wave duality of light is not an either/or phenomenon depending on the context. That lightwaves travel as waves, but also carry secret separate autonomous photons that we cannot see. De Broglie's wave theories have been largely dismissed by contemporary physics in the context of how light behaves and propagates.

It is not denied that strange things can happen in the world of subatomic particles that we still do not understand. But to date none of the aforementioned phenomena has been credibly proven experimentally. These examples among others reflect the very dubious nature of quantum physics and why many scientists think that today's 21st century quantum physics is very overdue for a monumental change in a new direction. The first step in that new direction starts with understanding the true nature of light as postulated in this book.

*

Eaton's Constant

The German physicist Max Planck introduced his so-called *'Planck's Constant'* in 1900 as part of his efforts to formulate Planck's Radiation Law (also called 'Planck's Law). Planck's Radiation Law attempted to explain how light radiates from objects. For example, he postulated that when heat is applied to an object, its temperature rises and it begins to emit light.

Planck's Radiation Law was eventually shown to be mostly baseless. Nevertheless Planck made valuable contributions to quantum physics by inspiring further research that led on to other things in the study of subatomic particles.

Part of the problem with Planck's Radiation Law is its convoluted mathematical equations considered to be too long, difficult and implausible. Also, he made contradictory assumptions at different stages, as Albert Einstein has pointed out.

Regarding Planck's Constant, Max Planck was correct and ahead of his time in postulating that light is created or emitted as streams of separate photons rather than as waves of multiple joined-up photons. But Planck was incorrect to postulate his famous 'constant'. In fact Planck's constant is not constant.

Planck's constant can only ever give mathematical approximations, never a precise, everlasting mathematical calculation. For example, much effort was made trying to use Planck's constant to define a kilogram of weight so as not to rely on a lump of metal stored in a vault near Paris. Such an effort failed because Planck's constant can only ever give approximations. Eventually, after many years, Planck's constant had to be changed and locked into a different equation to make it work as a constant for defining the weight of a kilogram.

"The International Committee for Weights and Measures (CIPM) approved a redefinition of the Planck constant by adjusting the constant to be exactly $6.62607015 \times 10^{-34}$ kg·m2·s−1 rather than 6.626×10^{-34} joule-seconds. The new definition took effect in 2019" (source: abridged extract from Kilogram, Wikipedia.org).

"The physicists at the beginning of the 20th century over-estimated the Planck constant, and this gave rise to universal constants that do not exist in the natural world in itself" (source: Koshun Soto, et al, The Planck Constant Was Not a Universal Constant, Journal of Applied Mathematics and Physics, Vol.8 No.3, March 2020).

The reason that Planck's Constant can never be a constant is because it is born from spurious light-wave theory which postulates that light's energy is determined by the rate of the electromagnetic oscillations of photons. Planck's Constant is designed to calculate by how much a photon's energy increases when the photon's electromagnetic oscillations go up by one hertz. But a photon's energy is always the same, and its EM oscillations always remain the same, thus rendering Planck's Constant as baseless.

Planck's Constant has in effect been rendered as a meaningless equation, so here is a suggested replacement.

Eaton's Constant Explained

The constant speed of light is kept constant by the constant speed of its electromagnetic oscillations. The author would therefore humbly suggest that one electromagnetic (EM) oscillation of a photon would be a good reliable constant. This one oscillation of electromagnetic energy is regarded as the smallest possible unit of energy that can be measured.

"Photons are the smallest possible particles of electromagnetic energy and therefore also the smallest possible particles of light". (source: Photons, Office of Science, US Department of Energy, Washington, USA).

The value of Eaton's Constant is the energy of one photon, expressed in joules. In calculating this value the following seven points should be taken into account:

1. A single EM oscillation is the total energy of a single photon. There is no difference between the energy of one EM oscillation and the total energy of one photon.

2. A single photon oscillation refers to a single change from electricity to magnetism (or vice-versa).

3. A single photon cannot have and does not have a wavelength or a frequency. This is so because a wavelength is the distance between two moving photons. And the frequency is the concentration of photons in a given light ray. So any calculation of the value of Eaton's Constant cannot include any value of a wavelength or frequency in the calculations.

4. The amount of energy of a single photon is the amount of energy given to the photon by the electron that created the photon. All electrons create photons with the same amount of energy, and all photons are created equal. Hence Eaton's Constant is equal to the energy released by an electron to create one photon.

5. For the avoidance of ambiguity, Eaton's Constant is based on the hydrogen atom, and on a valence electron jumping from the second level to the third level to release a photon. The constant is expressed as electrovolts or electron volts of energy. Thus, Eaton's Constant is given as 3.02×10^{-19} Joules. This can also be interpreted as *"3.02×10 to the power of -19 Joules"*.

6. It is appreciated that when a valence electron absorbs the energy of a photon it will jump to a higher orbital level to release a new photon. Therefore, the energy absorbed by such an electron will always be the same amount of energy whatever orbital level it jumps to. Thus to avoid ambiguity, point 5 above

mentions jumping from level 2 to level 3 in the calculation of Eaton's Constant. But in fact the jump from level 2 to any other level can be used with the same mathematical result because the energy absorbed from a photon by any valence electron will always be the same. An electron will always release the same amount of energy absorbed from a photon.

7. The mathematics of Eaton's Constant in calculating '3.02 x 10^(-19) Joules' is the same mathematics that is normally used for calculating the energy absorbed by electrons in jumping energy levels. So the question is: If an electron in a hydrogen atom jumps from the second energy level to the third energy level, how many Joules of energy must the electron absorb? The answer is 3.02 x 10^(-19) Joules. The energy absorbed by any valence electron from any photon will always be equal to the energy of one photon.

Note: The spurious nature of light-wave theory described in this book states that electrons absorb different *wavelengths* of light rather than different *photons* of light. In other words, that different levels of light energy are absorbed and emitted by electrons, thus endowing photons with different levels of energy. This is regarded as baseless.

As mentioned, Planck's constant is ambiguous and baseless because it cannot be measured precisely, only approximately. However, one oscillation of a photon can be measured very precisely, and of course the energy of a single photon-oscillation (equivalent to the energy of one photon) is the same anywhere in the Universe.

The whole point of a mathematical constant is that it should serve as a precise, reliable 'real world' reference point that never changes. Hence, using the energy of one oscillation of a photon makes a perfect and simple constant, and it remains constant because it is 'tethered' to the universal constant speed of light.

At the time, Planck and those involved in light-wave theory could not entertain the energy of one photon as a constant because of their mistaken belief that the rate of photon oscillations determines the energy of photons, and that the energy of a photon can vary from photon to photon. Those beliefs led Planck to formulate his mistaken constant.

*

How is light-energy measured?

In a previous section of the book it was shown that the energy of a given beam of light or beam of electromagnetic radiation depends on the concentration or density of photons. The real physical distance between the moving photons determines the quantity of photons (how frequent they are) and hence the degree of energy. The more 'bunched up' the photons, the greater the energy. But how exactly is the degree or strength of energy expressed in contemporary physics?

There are various ways to calculate electromagnetic energy. To complicate things further, there are also various types of energy plus a variety of energy definitions such as kinetic, nuclear, potential, thermal, chemical, electrical, etc.

To keep things simple and within the scope of this book, we will limit ourselves to electromagnetic energy and how such energy is expressed and calculated in physics. A well-known formula used for this purpose is $E=hf$:

E: represents the energy of a given ray of light or group of photons, typically measured in Joules (J) per second.

h: is Planck's constant, approximately equal to 6.626×10^{-34} Joule-seconds (J·s).

f: represents the frequency of the electromagnetic radiation, measured in Hertz (Hz), which is cycles per second.

The energy of light can vary enormously, from radio waves to gamma rays, and is typically expressed as the amount of energy in a one second sampling.

Historically, this formula was first proposed by Max Planck in 1900 when he was attempting to calculate how much energy escaped from various objects (the blackbody experiments). Since then, scientists have tried to use the equation E=hf to calculate the amount of energy in a given beam of light or electromagnetic radiation. Such efforts have failed because, as already mentioned in the section 'Eaton's Constant', Planck's Constant (h) is not a true constant.

You will know from reading this book that the fallacy of Planck's Constant is that it postulates that the energy of a photon comes from the rate (speed) of the oscillations of a photon. Because of this fallacy, all attempts to use E=hf in modern physics have ended in failure unless the 'h' element of the equation is changed so as to make it more accurate.

There is no intention here to disparage Max Planck. Planck's Constant has served us well and has led onto other developments in physics. Today, anybody using the formula E=hf can consider alternatives to <<h>> for greater accuracy such as the Dirac constant, Boltzmann constant, Avogadro constant, or the kilogram constant. And of course Eaton's constant as described in this book.

As regards the frequency in the equation E=hf, this can be calculated as follows.

The wavelength is the distance between the moving photons in a given sample of light. This can be calculated roughly for a known colour in the visible spectrum. For example, if looking at blue water or a blue beach ball, you will know that the blue light reaching your eyes has a wavelength (distance between moving photons) of about 650 nm.

If the colour hue is a mix of different wavelengths reaching your eyes, then you take the mean of the different wavelengths. Then by knowing the wavelength, you calculate the frequency by dividing the speed of light by the wavelength.

In contemporary physics, the formula E=hf is no longer used for accurately measuring the energy of light or electromagnetic radiation. It is far easier and simpler to use modern computerised spectroscopy.

This works by analysing a sample of light or electromagnetism as it goes through a spectroscope. The light is then broken down to its constituent colours. The resulting colour hue then tells you the energy of the light or electromagnetism.

Remember that the colours tell you the distance between moving photons, and by knowing this distance the spectroscope gives you the concentration (frequency) of photons in the sample being analysed. The degree of photonic concentration determines the degree of energy.

*

What is the brightness of light?

The brightness of light depends on the energy of light. We take the words luminosity and brightness to mean the same, although semantically some would argue there is a difference depending on perception and distance from the source of light.

A question that is often asked is whether the brightness of cosmic objects is gradually diminishing. However far away from the object, the brightness of the light received on Earth stays the same on human time-scales assuming there is no significant change in the status of the object creating or emitting the light.

As light moves forward its light rays are continually expanding outward like the growing face of a cone. This means that when such light is received by an observer, the light will have become 'thinner'. It means that less light rays meet the observer. But as such light is continuously being produced (as if it were a continuous ray of light between a star and Earth, every time we dip into the ray to see it, it should not have changed or become 'thinner' than it already is (at least not on human time scales).

Another consideration is that sunlight and starlight are known to be very lethal to human beings because they are mostly gamma rays. But two things happen: Firstly, by the time the gamma rays reach Earth they have greatly thinned out because of the way all light spreads out as it travels. And secondly, because of light refraction any remaining gamma rays are destroyed by Earth's atmosphere.

The brightness or luminosity of light can vary if it is not coming from a distant cosmic body. As we all know, light becomes less bright the further away the source of light. The degree of brightness or luminosity depends on the frequency of such light. And the frequency depends on the concentration (density) of photons in such light.

Note that the brightness of light depends on the frequency, i.e. it depends on the concentration of photons in a beam of light. So the greater the density of photons in a light ray, the greater the brightness. The intensity of brightness or luminosity depends entirely on the light's frequency, and remember that the frequency depends entirely on the time-intervals (physical distances) between the travelling photons. In short, the density of photons determines the strength of brightness.

*

Why can't the speed of light change?

This is like asking why can't a person fly like a bird? It's not our nature to fly like a bird. Equally, it's not the nature of light to do anything but move at its given speed of light. It is simply the nature of electromagnetism to always move at the constant speed of light, never slowing down or speeding up or stopping and then going.

As discussed, in physics it is thought that light moves as a result of oscillating electromagnetism which creates the energy of movement. Given that light moves at a constant speed of c (unless something gets in its way), it follows that the rate of oscillating electromagnetism is the same for all photons. In other words the rate of oscillating electromagnetism (common to all photons) is what sets the constant speed of light everywhere.

Note: Don't confuse the oscillation of electromagnetism with the frequency of light. The rate of oscillation determines the universal speed c of light, thus all photons oscillate at the same rate. There is no relationship between the rate of oscillation and the frequency of light (remember the mentioned 'big misunderstanding').

As soon as light is created and emitted from a source, such as a light bulb, the photons immediately start their journey at the full speed of light (they don't start from zero and gradually build up speed). And they will never stop unless something gets in their way. Let's look at some examples of things getting in the way of light.

When light hits a metallic surface: when this happens the photons are absorbed into the atoms of the metal (the photon that went in is now gone). Then the electrons in such atoms emit new, different photons that replace the entering photons. So light went into the metal, and different light came out of the metal. As mentioned, note that light did not bounce off the metal (it was not reflected). Some dense metallic surfaces such as lead absorb light and fail to emit incident light. Colloquially, the surface of lead is said to be 'non-reflective'. Typically, metal is said to emit less than 70% of the light received.

We should be clear that light is never reflected under any circumstances, and it can never bounce off something, not even a mirror. But colloquially we often use the word 'reflection' in reference to the light we see around us. As mentioned, just about everything that we see is made possible to see because of incident light. That is, when light hits an object, it is absorbed into the object's atoms, and then the electrons of the atoms emit new (different) light known as incident light. As discussed, when we see a car, the image of the car that we see with our eyes is entirely derived from incident light.

Note: when light hits a surface at an angle, the incident light comes out at the same angle as shown in the following image:

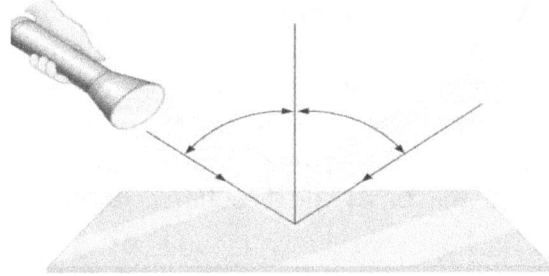

When light hits a mirror: when this happens it is similar to when light hits metal, except that with mirrors at least 95% of the light comes out again as incident light.

What happens when you look into a mirror? (Points A to E).

A. Let's assume you are looking into a bathroom mirror surrounded by daylight or artificial light. Given that you are surrounded by such light, many photons from the light source are being continually absorbed into every surface around you, and also into your face, and almost simultaneously new incident light is being emitted out of everything around you, and out of your face.

B. So you have incident photons being emitted from everything around you, and also being emitted from your face. If people didn't continuously emit incident light we wouldn't be able to see such people. This means the photons being emitted by your face go in straight lines to the mirror that you are looking at.

C. As these millions and millions of incident photons go from your face into the mirror, they are absorbed into the atoms of the mirror. And then the electrons in those atoms of the mirror emit new incident light in the form of incident photons that go out of the mirror in all directions, albeit in straight lines.

D. The following image, courtesy of explainstuff.com, shows how light rays (streams of incident photons) go out of your face and hit the mirror. Then those incident photons are 'reflected' back at the same angle. In this image the word 'reflected' is used colloquially with reference to absorption and emission of incident light.

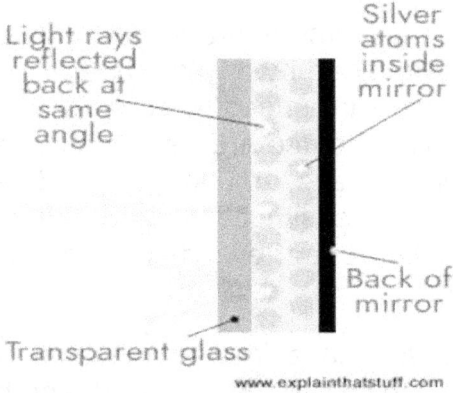

As you look into the mirror, the mirror will be continually emitting incident photons in all directions, albeit in straight lines. Some of those photons will end up going in straight lines from the mirror to your eyes.

E. As explained above, the incident photons going from the mirror to your eyes arrive at your eyes in a journey-time that is a little longer than the speed of light because of a longer time-interval between each incident photon. This longer time-interval is picked up by the eyes and brain and translated into a particular colour that corresponds with the mentioned time-interval.

As explained in the section *'How do we see colours?'*, each light ray (or group of photons) coming to your eyes carries a mix of physical distances between the moving photons in the light ray. Each distance is a wavelength, so the mix of wavelengths in a given light ray gives the brain a light recipe in the form of a heat signature for the particular colours you are seeing at that moment.

To finish on the subject of a mirror, the way that images can be seen in nature such as in water, ice, and other shiny surfaces is a similar process to the way images can be seen in a mirror. No light bouncing or light reflection is occurring in such phenomena. Here is an example:

When light hits a thin transparent pane of glass: Most of the light will go straight through the glass without being affected. This happens because the energy of the photons is not enough to excite the electrons in the atoms of the glass. But very roughly about 5% of light is indeed absorbed into the transparent glass and then sent out again as new incident light. This is why you will only get a very faint image of your face if you look at the transparent glass.

In summary, when we look around us and see a multitude of things (trees, cars, rooms, streets, people, etc) everything that we see is derived from incident light. That is, we are seeing light that has been absorbed and then emitted as brand-new incident light from everything around us.

It is estimated that when we are generally looking around, trillions of photons enter our eyes every second (about a quantity of 1 followed by 16 zeros). Many just dissipate without triggering any colours or images. This large input of photons is completely normal and innocuous, and it is how we have evolved as a species.

*

Does light have mass?

In classical physics mass is defined as matter, and matter is defined as anything made of atoms. So ultimately, mass is said to be made of atoms. But light is not made of atoms, only photons. So light is said to be massless.

There is much confusion on this subject, so here is further clarification. In contemporary physics it is clearly postulated that light has no mass. In other words, the photons that make up light are deemed to be massless particles. The reasoning is that the electromagnetic nature of photons is inherently massless by virtue of not being made of atoms. Is this so?

The answer is a firm yes, light is massless given that mass by definition is made of atoms. Some other kinds of subatomic particles are also massless in the sense of not having atoms.

When it comes to the energy of light it is a different matter. All photons carry electromagnetic energy, and such energy comes from its photons, not from anything to do with mass. The electromagnetic energy of light can be measured with great precision.

What about kinetic energy, i.e. the energy of movement? Light has no kinetic energy because it has no mass. So even though light moves, it has no kinetic energy arising from its movement. Kinetic energy and mass go hand-in-hand. You cannot have mass without kinetic energy, and vice-versa.

A big puzzle faced by Einsteinian relativity is the following: if light is massless, how can it be subject to gravity? The general theory of relativity is emphatic in stating that light is indeed subject to gravity.

The answer given by relativists is that the electromagnetic energy of light makes it fall prey to gravity and be pulled into space-time (the curvature of space). But there is no evidence for this and no credible explanation as to how exactly space-time exerts a pulling force on electromagnetism.

When one asks whether light has mass, the underlying issue is gravity. In physics, anything that has mass is subject to gravity. But however you may define mass, it is a fact that subatomic particles that may or may not be made of atoms are also subject to gravity. For example, electrons are not made of atoms yet they are subject to gravity. In fact, all known subatomic particles, whether or not defined as being massless, are subject to gravity, except light.

Light is the exception for one simple reason: the way it moves. When a photon moves, the inbuilt electromagnetism pushes the photon forward at right angles. So as a photon oscillates it is pushed forward at a perpendicular direction of propagation. As any physics textbook will tell you, each time the photon oscillates, the whole photon moves forward at right angles to the direction of energy transfer in a movement referred to as a transverse sinusoidal vibration. Hence, the photon can only ever move exactly in a straight line as depicted in the following image (and hence not fall prey to gravity):

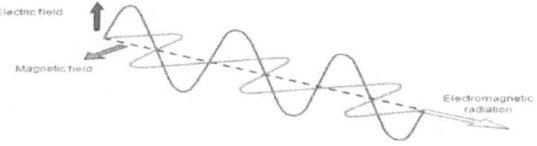

If light were to change direction or bend it would mean the light would have to move forward at some other angle, i.e. not at right angles to the direction of energy transfer. The light would no longer be moving as a transverse ray. There is no evidence that this is possible or that it ever happens.

This is why light can never change direction or bend, and this is why light can never fall prey to gravity. This simple fact means that light cannot follow the curvature of spacetime as it passes near a star. And it means that gravitational lensing, solar eclipses and other cosmic phenomena do not involve light bending - they involve light refraction. It means the theory of general relativity falls apart as being baseless.

As to why light spreads out in all directions, albeit in straight lines, this occurs because all light is born from non-stop moving electrons spinning around the nucleus of the atom. Hence, photons are emitted while electrons are on the move.

*

Can light ever bend?

Light can never bend under any circumstances as explained in the previous section. Nor can it go round corners or change its angle of travel. If you're in an L-shaped room and someone shines a torch from around the corner, you will see the light as a result of the light's attenuation from all the surrounding walls and surfaces (no light bending has occurred).

Three examples are given in what follows: (a) fibre optic cables, (b) refraction, and (c) cosmic light bending myth.

(a) Fibre-optic cables. When light follows the twists and turns of a fibre-optic cable it never bends or changes its angle of travel. Here is an explanation.

The following image, courtesy of howstuffworks.com, shows a fibre-optic cable:

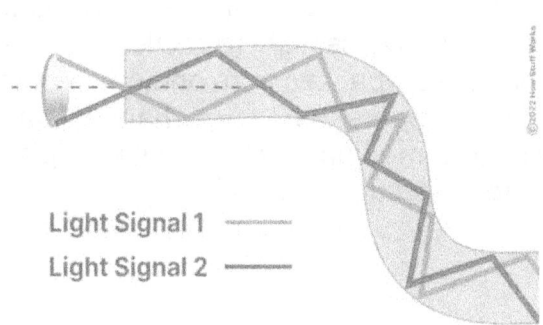

We will limit ourselves to just explaining how light travels along a fibre-optic cable, rather than explaining the complexities of fibre-optics. Take for example light signal 2 in the above image. When light is fed into the cable it goes in a straight line and hits the inner side of the cable. Such light is then absorbed into the cable and new incident light is emitted outward (inside the cable) at the same angle. This new incident light also goes in a straight line and hits the inner side of the cable again further down, and so on as shown in the image. This is how light follows the bending turns of the cable without ever actually bending.

You will appreciate that each absorption and emission will slow the light down a little. Hence, the overall speed (journey-time) of light going through a fibre-optic cable is a little slower than the speed of light.

(b) Refraction. Previously we discussed incident light. The phenomenon of incident light is caused by refraction. In physics, light refraction is the redirection of a photon as it passes from one medium to another. Thus, light appears to bend or change its angle when it goes into a medium such as water or glass. Here's an image to illustrate this:

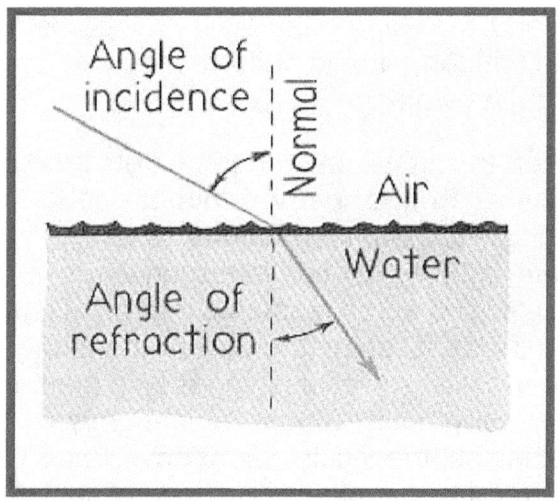

In the above image, when light hits a medium such as water it appears to change its angle of travel due to refraction. This phenomenon is well known to science. The refraction is caused by the light being absorbed into the electrons inside water molecules as soon as it hits the water, and then being emitted as new (different) incident light at a changed angle. This gives the illusion that as the light entered the water the *same* light changed its angle.

As water is a dense medium, the light's journey-time through the water will slow down. The change in speed is caused by absorption and then emission of incident light from the atoms in water molecules. What happens is that when light hits water it is absorbed into the atoms of water molecules. And when incident light is emitted from the molecules, this incident light travels on to the next molecule in its path, and then again the light is absorbed, incident light is emitted, and so on from molecule to molecule through the water.

When the light is absorbed, the electron inside the molecule **allegorically** rotates 180 degrees (technical explanation follows shortly) and emits incident light which continues onwards to the next molecule, and so on, ensuring a straight line of travel for the incident light in the water. The empty distance between water molecules varies with their movement but is very little (about a nanometre or less). Nonetheless, the incident light travels in a vacuum to the next molecule at the full speed of light.

Thus, when light travels from molecule to molecule through water, it's not the same light. Each time light is emitted from electrons inside water molecules, it's different light, and incredibly, the electrons do an about turn and send the new light in the same overall direction of travel. You have to marvel at the wondrous nature of light.

So when incident light travels to the next water molecule it does so at the full speed of light c, but its overall journey-time of light going through water from A to B is longer compared to the speed of light going the same distance from A to B unencumbered by a medium. It is only the attenuation (absorption/emission) of light that slows it down. Thus, it is indeed accurate to say that light travels at speed 'c' *in any kind of medium* (not just in a vacuum) until it is stopped or absorbed.

The photons of light always move at their constant speed 'c', whether or not in a medium such as water

Don't confuse the speed of light with the journey-time of light. The former refers to the speed of individual photons at speed c. The latter refers to the time it takes a whole given bunch of photons to travel from A to B.

Why does light move in a straight line under water?

The answer is gravity as fully explained in Snell's Law. Electrons are subject to gravity but not light. When an electron absorbs a photon, the over-energised electron will release a new photon. But in doing so, the photon is released at an angle corresponding to the 'normal force'. This normal force (or just 'normal' in physics) is governed by gravity. Put simply, the normal exerts a force of gravity that affects the direction of emission of the photon.

As photons are not subject to gravity, when they are emitted by electrons they continue in a straight line in the direction set by the electron's emission. But as water moves, changes temperature, and may contain impurities, the line of travel of light as a whole through water (from start to finish) will normally not be entirely straight.

So in the case of light going through water, when an electron emits a photon, the photon is emitted in the same direction as the incoming photon that was absorbed. Technically, light continues straight on in water after entering it because the angle of incidence is 0 degrees, which means the angle of refraction is also 0 degrees. This is fully explained in Wikipedia.org under 'normal force' and also under 'Snell's law', so the well-known and fully verified technicalities of the normal force will not be repeated here.

But light-wave theory has a different explanation which goes like this: All light travels as light-waves (fields of joined photons). So in water, electrons emit whole light-waves rather than individual photons. These light-waves do not travel in straight lines inside water, but they appear to do so. Why so? Because when light-waves enter water their wavelengths become much smaller, i.e. their electromagnetic oscillations slow down, giving the appearance of going in straight lines.

Light-wave theory does not deny Snell's Law of refraction. Rather, a different version of Snell's Law is postulated that is derived from Huygen's wave theory. This states that each point on a wavefront is a source of secondary waves. Huygens wave theory is largely dismissed in contemporary physics because the theory cannot explain how light diffraction or polarisation occur. Huygens wave theory remains a foundational concept of light-wave theory, while also remaining unproven and baseless except for those entrenched in misguided light-wave theory. Whereas the veracity of Snell's Law and the 'normal force' of gravity has been verified in countless experiments.

(c) Cosmic light bending myth. You will know from reading this book that light always travels in straight lines through the cosmos, albeit in a vibrating ellipsoid movement. As discussed in our sister book the *Final Theory of Everything*, the phenomenon known as 'Gravitational Lensing' is in fact caused by refraction rather than by gravity. Even massive objects like a cluster of galaxies will not make light bend or change direction as it passes by, as evidenced by many astronomical observations.

But what about when astronomers see the same star in two different places simultaneously in the night sky? Doesn't this show that such starlight must be bending into a different direction? Consider the following image:

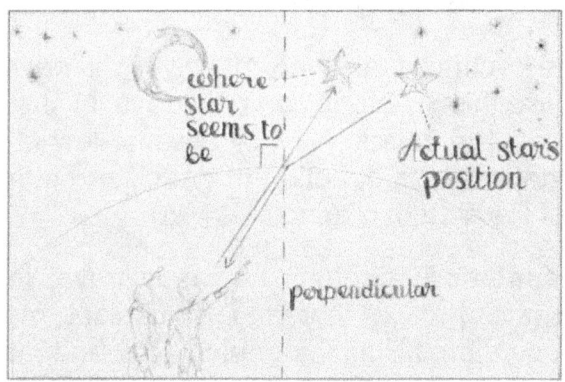

In the above image the same star appears to be in two different locations simultaneously. How can this be explained if light cannot bend or change direction? This is what happens. As the light from the star's actual position approaches Earth it hits Earth's atmosphere. This is akin to hitting water.

When starlight hits the atmosphere, it is absorbed and emitted as new (different) incident light. But this refracted light is emitted at a more downward angle towards the 'normal' (the perpendicular dotted line in the image). This is akin to when light hits water and changes its angle downward. This means the refracted starlight coming to your eyes is new (changed) light, it is not the same starlight that has simply changed its angle of travel.

So when we see the arriving starlight, it may look as if it is coming from another location in the sky (see image) but this is just an optical illusion. No cosmic light bending, or change of direction through gravity, has occurred at any time.

The technical answer to *why light never bends* **goes like this:** Photons can only travel in straight lines because the rotational flux always lies in a plane orthogonal to the direction of travel, and therefore no part of the field can bend.

Translation: photons can only travel in straight lines (albeit in a vibrating **sinusoidal movement**) because the oscillations of electricity and magnetism are at right angles to the direction of movement. Hence no part of the photon's field can bend when moving. This 'leapfrogging' between electricity and magnetism also ensures the speed of light is kept constant without slowing down or speeding up.

Countless experiments and research projects show that light can only travel in straight lines. If light were affected by gravity it would show up in such research as a non-straight line, but this is not the case. *"Not only does light travel straight, it travels in straight lines through a given medium. Various first-hand sources of evidence point to light's linear path"* (source: Light Physics Narrative for 11-14, travelling in straight lines, spark.iop.org).

*

Does light carry information?

Yes, a light ray does indeed carry information. Photons themselves do not carry information, but a stream of photons does carry information which can be gleaned from the frequency of the light ray. The frequency (i.e. the time-intervals between moving photons) determines the colour spectrum applicable to a stream of photons. And the colour spectrum in turn can reveal information about certain chemicals, elements and more. Here's an example:

Let's suppose we are receiving light from planet X out in space. And let's suppose that the frequency of the light coming from planet X indicates the presence of silicon. This means that the incident photons received from planet X into the spectroscope on Earth displayed a spectrum of colours corresponding to silicon.

As mentioned, incident light travels at the same speed of light, but the time-intervals between each moving incident photon determines the frequency. So when the starlight (local sunlight) was absorbed into silicon locations on planet X, the starlight was absorbed and incident light was emitted out of planet X that went to Earth. The time-intervals between the photons of this incident light are what provide a spectrum of colour relating to silicon.

So to summarise, the time-intervals (physical distances) between the moving photons arriving from planet X provide a given frequency of the light rays received. This frequency can then be checked against lists of colour spectrums and matched with their corresponding elements and chemicals. A detailed study of the spectra from planet X may also reveal whether it has an atmosphere and other elements, apart from silicon.

In this example no information has been carried by the photons themselves. But much information can be gleaned from a stream of incident light by using powerful computer-aided spectroscopes. We can only do this with incident light. If the light received comes directly from the source that created the light (with nothing getting in its way), spectroscopes on Earth will not be able to determine any colour spectrum for the light. This is because such light (i.e. unpolarized light) provides the whole colour spectrum (colour white), so no information can be extracted.

However, even this is not entirely so. Some astronomers and physicists are now discovering new ways to use light spectrums to analyse light. They have discovered that light from any source, whether a candle or a star, is composed of a combination of wavelengths (frequencies) depending on what atoms and molecules are emitting the light. This science (spectroscopy) allows astronomers to determine what elements may be present in a given star. So it turns out that light created by a star can include a mix of refracted light (i.e. incident light) and unpolarized white light that can reveal information about its source.

When we go about our daily lives everything around us is continually emitting many millions and trillions of incident light photons in all directions. But we don't see or receive most of it. We only see some of it. That is, we only see things when some of those streams of incident light meet our eyes.

As an example, think that you are standing in a shopping centre. You gradually turn around looking at everything. At any one moment in time you will be looking at something. At that precise moment many different incident streams of photons will be coming at you from the objects that you are looking at, and such streams will be going into your eyes.

Equally, every part of your body and clothes will be continuously emitting trillions of incident streams of light in all directions. If this were not so, nobody could see you.

The key point to understand is this: your eyes will be receiving many different streams of photons, and each stream will have a different mix of wavelengths that together provide the brain with a particular light recipe. Those light recipes will at that moment make us see the panorama of colour and shapes being looked at.

*

Virtual Video Camera

Given the fundamental nature of light as revealed in this book and given the current level of spectroscopy in science, the following prediction is offered as a matter of speculation:

It is predicted that by the year 2030 or perhaps sooner it will be possible to see close-up moving video images in full colour and sound, coming from the surface of any planet or star, however far away they may be. Here is how this will be achieved and lead to a scientific revolution in cosmology (eleven points follow):

1. A story to tell. Thousands of exoplanets have been identified and each one will have its own story. A unique way to receive incident light from such exoplanets is revealed below. With that we will be able to explore the cosmos in an extraordinary way like never before.

2. A wealth of information. Each of the many millions of incident rays arriving on Earth will have a different frequency depending on where the incident light is emitted from. Many different incident rays may come from the same planet or star, thus providing a wealth of information.

3. Artificial intelligence. Future spectroscopy will become much more advanced on the back of more powerful computers and artificial intelligence. Thus, such spectroscopes will be combined with very powerful computing and AI capabilities that will be capable of analysing and compiling the many different frequencies of incident light received from just about any chosen planet or star.

"AI systems provided one of the most promising potential developments for the future of analytical molecular spectroscopy, noting that AI systems presented a more dynamic analytical solution set" (source: Artificial Intelligence in Analytical Spectroscopy, Howard Mark, et al, Spectroscopy, vol 38, issue 6, June 2023).

4. Virtual video camera. Given future developments in spectroscopy, it is theoretically and technically probable that the millions of different light frequencies received will be converted into a panorama of full colour, just as we humans are doing every day when we look at things on Earth. This means that through the virtual video camera we will be able to see things on the distant planet in full colour and shape, just as if we were standing on the given planet looking around at things. It will be a virtual video camera that will be possible to 'put' on the surface of any planet or star that sends us incident light.

5. Just like a movie. The many millions of incident light rays received by future spectroscopy (as described in this prediction) can then be compiled or converted into moving images. This is a well-understood science given current digital cinematography and video technology. It will be a small step to apply such knowledge to the many millions of images extracted from incident light received from a given planet. Let us remember that human eyes can already process millions of different light frequencies to end up seeing the full colour and movement of things around us. We are nearly at the point of being able to do the same with powerful spectroscopy as it develops in the future.

6. Stars and planets. Thousands of exoplanets have been identified by astronomers and they all emit incident light arriving at Earth. But much of this incident light is drowned out by the planets' local stars. However, the detection rate of such incident light is greatly improving as technology improves. There will be no shortage of distant planets that we can explore, using advanced spectroscopy that will give us full high-definition video images of such planets. The same goes for starlight - incident light from stars can be analysed the same way as incident light from planets, showing the composition, age and other factors of a star.

For high quality definition, free of distortion from the Earth's atmosphere, relay stations can and will be set up in the form of one or more satellites. Incident light from a given planet or star could then be received by a satellite relay station and beamed to Earth free of atmospheric distortion.

7. No distance-barrier. Putting all this together, it is predicted that by 2030 or before, we will be able to see full video images as if we had a video camera stationed on a given planet. If the incident light from a planet is, say 100 light years away, it means we will be looking at video scenes that occurred on that planet a 100 years ago. Distance will be no barrier because the incident light from that far-away planet is already continuously arriving on Earth.

8. Adding sound. It is also predicted that it will be possible to add sound to the mentioned video images. It is not suggested that incident light will somehow be able to record and carry audio waves to Earth. But using the ingenuity of AI and very powerful computer technology it will become possible to incorporate sound into the silent video images, giving a very realistic sound-track to such videos. This will give us a full virtual video, coming to Earth from distant planets however far away.

Adding sound to silent movies for example can already be done, and the groundwork for adding sound to endeavours in astronomy is well underway. *"Astronomy is often thought of as a visual science that produces stunning images of the cosmos, but it's possible to hear it as well"* (source: Patchen Barss, How sound is providing new clues about the Universe, Oct 2023, bbc.com).

Advanced technology will be able to take clues from the silent video of a far-away planet, such as dust movements, changing scenery, weather, and any kind of movement, so as to add a realistic sound-track to the video.

9. A revolution in cosmic discovery. Future spectroscopy technology as predicted here will revolutionise astronomy in unforeseen ways and tell us things about other planets and stars that we cannot do today, even with super-powerful telescopes. The search for extraterrestrial life and the way other planets have evolved are just two of the many things that come to mind.

10. A full colour audio-visual movie. High quality optic resolution of the virtual video images will be achieved with artificial intelligence and powerful video enhancing software which will be capable of greatly improving the resolution of such images. Thus, the end result will be high quality, full colour & sound videos (just like a movie) of distant planets and stars as if we had put a physical movie camera on the surface of a planet or star.

11. A waiting laboratory. We humans are lucky to have a solar system with lots of planets nearby. It is estimated that many stars have no planets, or one to two planets at most. Less than 20% of stars have three or more planets.

Our solar system is a waiting laboratory that can be used now to develop, test, and fine tune a virtual video camera. The solar system gives Earth an abundance of cosmic incident light. By experimenting with incident light coming from the moon and nearby planets, scientists can today start using the latest generation of computerised spectroscopy and be ready for putting virtual video cameras on faraway exoplanets.

In a sense, the predicted virtual video cameras already exist and are already in place on planets and stars, and we are already receiving their video images on Earth. But we humans do not yet have the capability to detect and extract these video images from light rays. For example if a planet is 50 light years away, we will be able to 'put' a virtual video camera on that faraway planet with no time delay at all because the light rays arriving on Earth already 'contain' such video recordings, albeit they will show us things 50 years old in this example.

The use of spectroscopy in astronomy is a well-trodden path and that is why it is not fanciful to predict the development of virtual video cameras as postulated in this book:

"Astronomical spectroscopy is the study of astronomy using the techniques of spectroscopy to measure the spectrum of electromagnetic radiation, including visible light, ultraviolet, X-ray, infrared and radio waves that radiate from stars and other celestial objects. A stellar [colour] spectrum can reveal many properties of stars, such as their chemical composition, temperature, density, mass, distance and luminosity. Spectroscopy is also used to study the physical properties of many other types of celestial objects such as planets, nebulae, galaxies, and active galactic nuclei" (source: Astronomical spectroscopy, Wikipedia.org).

Scientists are urged to take up the challenge of developing advanced spectroscopy with the aim of giving the world virtual video cameras as proposed here. This has the potential to revolutionise our understanding and exploration of the cosmos, probably more than anything else.

*

Finding Extraterrestrials

This section reveals how scientists can find extraterrestrial life on other planets in a way that will revolutionise and greatly increase the success of SETI (Search for Extra-Terrestrial-Intelligence).

Light from planets in other solar systems can indeed be detected, but it's extremely challenging. Most of the planets beyond our solar system, known as exoplanets, are too faint and close to their much brighter host stars for direct observation. Instead, astronomers use indirect methods such as the transit method or the radial velocity method to detect the presence of exoplanets. The transit method involves observing the slight dimming of a star's light as an exoplanet passes in front of it, while the radial velocity method measures the tiny wobbles in a star's motion caused by the gravitational effect of an orbiting planet. These methods have allowed scientists to detect thousands of exoplanets, but seeing or detecting the light from these planets is still a significant technological challenge.

Even the best telescopes available are not able to see or detect any significant light coming from exoplanets. But there is a workaround called astrometry. This involves tracking the motion of a star using precise measurements. Using astrometry, exoplanets can be found by measuring tiny changes in the star's position as it wobbles around its centre of mass. The difference between the radial velocity and astrometry measurements is how scientists can discover the existence of exoplanets.

Planets are much less massive than stars, but they still exert gravity. When planets affect stars gravitationally, they cause the stars to move (wobble) slightly. Technically, a star's centre of mass called a barycentre is slightly affected by an orbiting planet. Sensitive instruments on Earth can measure this wobble movement very precisely, and astronomers can infer that one or more planets must be orbiting their star.

There is no shortage of exoplanets waiting to be discovered by scientists using the mentioned astrometry method. On average, it is estimated that there is at least one planet for every star in the galaxy. That means there's something on the order of billions of planets in our galaxy alone, many in Earth's size range. So we live in exciting times with every chance of discovering extraterrestrial life, whether it be sentient or not.

"The most exciting development in astronomy over the next 50 years is likely to be the study of exoplanets and whether there are signs of life beyond our solar system. Almost every star has a planetary system and that's a fantastic thing to investigate" (source: Mike Edmunds, Astrophysicist, President of the Royal Astronomical Society, The Life Scientific, BBC, April, 2024).

But there is a big problem facing cosmologists. Even though astrometry allows scientists to discover exoplanets, what then? How can we find out if any given exoplanet has any kind of life? This seemingly intractable problem has greatly prevented the successful search for extraterrestrial life. And it has greatly prevented SETI from being successful. If we could somehow determine the probability of life on any given exoplanet, it would then be possible for cosmologists (and SETI) to focus on such an exoplanet.

By greatly narrowing the search it would hugely increase any chances of finding extraterrestrial life. So given that scientists have found thousands of exoplanets, how can we 'shortlist' the exoplanets showing the greatest promise of having life? This will now be revealed in what follows.

The image below shows 'transmission spectroscopy' in action. When a planet goes past a star facing towards earth it means the planet is 'between' its star and planet Earth at that moment in time. But that *'moment in time'* can last for months. Remember that it takes Earth six months to go halfway round the Sun:

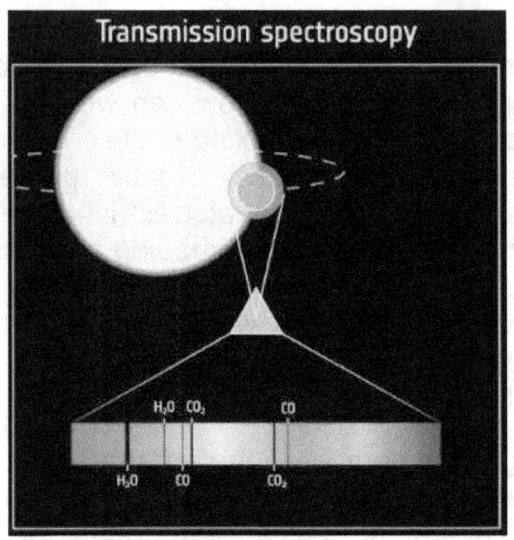

So while an exoplanet is going around its local star, and is positioned 'between' the star and Earth, scientists will have ample time to apply the 'Finding ExtraTerrestrials method' (FET method) described below. The FET method involves the use of a virtual video camera as described in the previous section and the six-step process that follows.

The FET method (F**inding **E**xtra**T**errestrials method)**

Step 1. Determine the exoplanet. Using well-established astrometry, the first step is to choose a given exoplanet having determined its presence around a particular star.

Step 2. Capture the starlight. The next step is to capture the star's light coming to Earth and send it into a spectroscope. If the star can be seen through telescopes on Earth it means the starlight has been arriving on Earth continually for a very long time. As explained in the previous section, if the starlight is received through a satellite relay station, then all the better because atmospheric interference on Earth will be avoided.

Step 3. Separation of incident light. Step three is to analyse the starlight using the mentioned advanced spectroscopy. This will filter out actual starlight so that you are left with just any incident light from the exoplanet for a more detailed analysis.

Starlight is mostly unpolarized (incoherent) light that is well recognized and understood by science. Whereas incident light is mostly polarised light. Therefore a sophisticated computerised spectroscope could quite easily filter out (disregard) unpolarised light, and focus on polarised light (incident light). The technology to do this already exists:

"A spatial light modulator can be used to modulate ... the incident light beam. Thanks to diffraction, the spectra from the coherent and incoherent part are spatially separated at the back-focal plane of a lens" (source: Xiang Li, et al, Separation of coherent and incoherent light by using optical vortex via spatial mode projection, Optics Communications, Volume 527, 2023).

Step 4. Analyse the incident light. The mentioned incident light will have been created by diffracted light coming to Earth from behind the exoplanet. You will know from reading this book that diffracted light is light that comes from behind an object, and then goes over the edge, horizon or lip and onwards. This is how we see an eclipse for example: the sun's light comes to you from behind the moon by being diffracted around the edges of the moon. Here again is an image of diffraction:

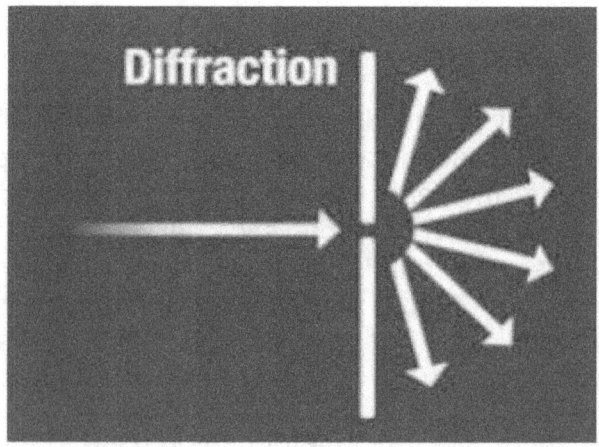

The phenomenon of light diffraction is well known in optic physics: any photons hitting the horizon, edge or lip of an object will be absorbed and then emitted outward as incident light in all directions over the horizon or edge. No light bending or changing direction has occurred. With diffraction the photons shoot out in all directions, always in straight lines, as explained by Snell's Law.

The following image shows diffraction in action. If diffraction was not a fact we wouldn't be able to see the incident light coming to our eyes from behind and over the edge/horizon of an object:

So coming back to our exoplanet, starlight will hit the orbiting exoplanet, but from behind from our perspective. Then some of that light will diffract around and over the circular edge of the planet and continue on its way towards Earth as incident light. If viewed by a powerful telescope, such incident light will no doubt be drowned out by the starlight that is simultaneously also coming to earth. So we receive a mix of starlight and incident light, providing the opportunity to filter out the unpolarized starlight so as to end up with just polarised (incident) light from the exoplanet.

Step 5. Distance no barrier. This mix of starlight and incident light would have been arriving on Earth continuously for many years. So when a given exoplanet is chosen as the first step in the FET Method, we don't have to wait for the exoplanet's incident light to arrive on Earth - it is already arriving continuously. We simply dip into the continuous light ray every so often until our spectroscope detects an element of incident light mixed in with the starlight.

If the light ray has no incident light it means that the exoplanet was not on 'this side' of the star when such light was emitted (this of course will be the case 50% of the time). If incident light is indeed detected it means the exoplanet was on this side of the star when the incident light rays started their journey towards Earth.

The mentioned incident light may have taken, say, 100 light years to reach Earth. But the incident light will be giving us a virtual-video-camera recording with no time delay, albeit showing us the exoplanet as it was a hundred years ago. This means that the distance of the exoplanet is no barrier; it could be a thousand light years away but as its light is already arriving on Earth it means the virtual video camera of that distant exoplanet is already available.

Step 6. Deploy the camera. The final step of this FET method is to 'put' a virtual video camera on the chosen exoplanet to give us video recordings as if there was an actual physical video camera stationed on the surface of the exoplanet. This will tell us, more than anything else, whether there is any sign of extraterrestrial life (sentient or not).

This 6-step FET method is destined to completely revolutionise humanity's exploration of the Universe and greatly increase the chances of finding extraterrestrial life. And who knows, it could be that some highly intelligent extraterrestrials, thousands of light-years away, already have their virtual video camera deployed on our planet, giving them a view of Earth as it was thousands of years ago.

As and when this FET method reveals good signs of biological life on an exoplanet that is, say one thousand light years away, it means that today (one thousand years later) that exoplanet could be harbouring sentient life. This would provide SETI, for example, with a good clue for a focused search of extraterrestrial life and a greater chance of success.

Millions of exoplanets are already beaming their incident light to Earth, and of these millions, astronomers have already identified thousands of specific exoplanets by mostly using well-established astrometry. By applying this FET Method, it means we have many thousands of exoplanets waiting to be explored by using virtual video cameras.

As mentioned, the video 'recordings' of such exoplanets are already continually arriving on Earth as light rays. But our level of spectroscopy science is not yet quite capable of grabbing those virtual video recordings that are already 'built into' the incident light of the rays. One day soon, scientists will be able to show us full video recordings (with sound, movement and colour) for just about any chosen exoplanet however far away.

Unfortunately, the big misunderstanding of light described in this book has only served to impede the advancement of cosmic exploration and the search for extraterrestrial life. This misunderstanding, which is deeply rooted in contemporary physics, drives much scientific research down blind alleys and dead ends. It is hoped that this book will help to redress this situation.

For example, you may well be wondering why hasn't a virtual video camera already been proposed or developed? The answer is that Einstenian general relativity (GR) is still very entrenched in contemporary physics. GR postulates that light falls prey to gravity. Thus, the phenomenon of diffraction as mentioned in step 4 above would not occur in GR. In brief, with GR it would mean that the starlight hitting an exoplanet from behind and then curving around the exoplanet to go onwards toward Earth would not be incident light (it would be the same starlight that followed a curved path to reach Earth).

In this GR scenario no virtual video camera would be feasible because no incident light from the exoplanet would be reaching Earth. This is a good example of how relativity is holding back science and why the feasibility or concept of a virtual video camera has not been considered.

The take home message: Light is an awe-inspiring and wondrous phenomenon that is at the heart of creation and the Universe as we know it. Light travels at 300 million metres a second, always in straight lines. It never changes speed, never bends or changes direction, and never bounces or reflects off things. When light hits anything it is changed or destroyed, and new light is emitted to replace the lost light.

Light can tell us about the composition of stars and planets among its many incredible feats, and one day soon light will help us discover extraterrestrial life. It is hoped that this book will help galvanise our understanding of the cosmos like never before.

*

Please see the next page for a message from the author →

Message from the author

Thank you for reading the *Final Theory of Light*. If you liked the book please leave a brief review at the site of your ebook or physical book purchase. Any other feedback is much appreciated as it will help improve and update future editions - for this, my email is: mailto@deliveredonline.com (please put **only the title** of the book in the email subject heading to make sure I see it). To tell others about this book they may go to www.deliveredonline.com or simply search the title on the internet.

Russell Eaton, author.

(See next page for another book by the same author)

*

Here is another book by the same author: *Final Theory Of Everything,* available as an ebook or physical book from most online and offline book retailers worldwide.

Final Theory of Everything

The *Final Theory of Everything* (FTOE) reveals for the first time a new 'theory of everything' to explain how all aspects of the universe are linked together, and why the universe is the way it is.

The holy grail of cosmologists has been to find a master theory that provides a singular, all-encompassing, coherent theoretical framework of physics that fully explains and links together all aspects of the universe. This book reveals precisely that: a grand unification theory that brings together the four forces of nature (gravity, electromagnetism, weak and strong forces) into one single force.

Along the way, many spurious concepts and misconceptions about the Universe are busted wide open, made possible by the *Final Theory of Everything*. For example, the book reveals why dark energy and dark matter are non-existent and unnecessary in the Universe. Other mysteries are resolved such as what keeps galaxies together, what's at the bottom of black holes, exactly what causes gravity, and at long last the true nature of quantum gravity is revealed.

By knowing how gravity works for the big and the small, a final unifying theory is revealed which is destined to dramatically transform astrophysics. It will not mean the end of today's physics, but it will certainly change a lot of things in science generally and lead to exciting new discoveries in the Universe.

This book is written for a general audience and for scientists & physicists - for anybody wanting to know more about our astonishing Universe and the world we inhabit.

The *Final Theory of Everything* by Russell Eaton is available as an ebook or physical book from most book retailers worldwide. For more information please visit: www.deliveredonline.com or simply do a title search on the internet.

*

(See next page for author bio)

Author Bio

Russell Eaton is British and the author of several non-fiction books, mostly relating to health and well-being. With a passionate interest in cosmology, his books 'Final Theory of Everything' and 'Final Theory of Light' are his biggest projects.

He has lived in the UK and in Ecuador, splitting his time between the two countries, and sometimes getting into a pickle because of it. Widely travelled, Russell Eaton keeps an interest in any and all the wonders of the world and the Universe.

He says *"We must always endeavour to eradicate bigotry and prejudice from science, and we must always be on guard when such pernicious influences come knocking at the door".*

***

www.ingramcontent.com/pod-product-compliance
Lightning Source LLC
LaVergne TN
LVHW020927090426
835512LV00020B/3246